"Cobb and Fetterolf have written a chemistry primer for the mature, inquisitive nonchemist. The book is rich with chemical history and applications to everyday life and includes principles of chemistry illustrated by experiments and demonstrations using household items. *The Joy of Chemistry* would make a wonderful supplement for chemical educators and also stands on its own as the book for the chemistry enthusiast."

—Stephen L. Crump, PhD
Savannah River National Laboratory,
Westinghouse Savannah River Company

"If your high school or college chemistry course did not fill you with an appreciation of chemistry in our everyday world, this may be the guide that you need to revisit the issue. Try a few of these experiments and read the discussion and you will recognize chemistry happening all around you."

—Don Franceschetti, PhD
Professor of Physics and Chemistry, University of Memphis

"An enjoyable read for both the scientist and nonscientist alike."

—Dr. Colin Moore
Research Fellow, Crompton Corporation

# THE JOY OF CHEMISTRY

THE JOY OF

# CHEMISTRY

## The Amazing Science of Familiar Things


CATHY COBB &

MONTY L. FETTEROLF

Prometheus Books

59 John Glenn Drive

Amherst, New York 14228-2119

Published 2010 by Prometheus Books



Inquiries should be addressed to
Prometheus Books
59 John Glenn Drive
Amherst, New York 14228–2119
VOICE: 716–691–0133, ext. 210
FAX: 716–691–0137
WWW.PROMETHEUSBOOKS.COM

14 13 12 11 5 4 3

Library of Congress Cataloging-in-Publication Data

Cobb, Cathy.
The joy of chemistry : the amazing science of familiar things / Cathy Cobb and Monty L. Fetterolf.
p. cm.
Includes bibliographical references.
ISBN 978–1–59102–771–3
1. Chemistry—Popular works. I. Fetterolf, Monty L. II. Title.

QD35.C56 2005
540—dc22

2004020144

Printed in the United States on acid-free paper

# CONTENTS

# PART 2

# ACKNOWLEDGMENTS

We are deeply grateful to our colleague Dr. Jack G. Goldsmith for his thorough review and excellent suggestions. Thank you so much, Jack, for being so generous with your time and helping us identify and rectify our many BHM. We also thank Francesca Pataro for reading long sentences and gently advising how to abbreviate them and for befriending our children while their parents were distracted. We extend our heartfelt thanks to Dr. and Mrs. Cobb for their patient assistance in reading yet another mound of manuscript and for offering, as ever, wonderfully insightful comments and suggestions. We thank Judy Dailey for providing excellent and perceptive commentary, and we thank Linda Regan, as always, for her guidance, without which none of this would be possible.

We thank Heather Ammermuller and Chris Kramer and all the production staff at Prometheus. Your friendly, gracious support and advice made all the difference.

We would like to extend a special thanks to our children, Mathew, Benjamin, and Daniel Fetterolf, for being willing to share their home with sometimes smelly experiments, frequently fussy and fretful parents, and a pile of papers where a living room should be. You helped with demonstrations, ideas, artwork, troubleshooting, and just plain patience, and we love you for this and so much more.

We would like to thank our friends and colleagues at Aiken Preparatory School and the University of South Carolina Aiken for their help and encouragement. We gratefully acknowledge Laura Bacon of Kennedy Middle School for introducing us to oobleck, and we thank Beth Beckham for her kind encouragement of all our efforts. We would also like to thank True Value Hardware of Aiken. They didn't know they were assisting with a chemistry book, but we couldn't have done it without their wonderful, extensive inventory and their kind willingness to track down our strange requests.

We would like to thank our students for being receptive sounding boards and we thank Joanne Morton and Linda Muse for being the best kinds of partners—friends.

We dedicate this work, with all of our love, to Neda Jo Fetterolf, a model of strength, resilience, and courage.

*See the blind beggar dance, the cripple sing,*
*The sot a hero, lunatic a king;*
*The starving chemist in his golden views*
*Supremely blest, the poet in his muse.*

—Alexander Pope, *An Essay on Man,* 1734

*The sight of the planet through a telescope is worth all the course on astronomy: the shock of the electric spark in the elbow outvalues all the theories; the taste of the nitrous oxide, the firing of an artificial volcano, are better than volumes of chemistry.*

—Ralph Waldo Emerson, Essays: Second Series, 1844

# APOLOGIA

Chemistry books usually fall into one of two categories—textbooks or children's books—which, in essence, excludes the general reader. There are some excellent works that address this deficit, including *Chemistry Connections: The Chemical Basis of Everyday Phenomena* by Kerry K. Karukstis and Gerald R. Van Hecke and *The Genie in the Bottle* by Joe Schwarcz.[1] Still, we have chosen to add our own effort to the mix, believing it to represent a slightly different approach. We have attempted to write a fireside chemistry: a chemistry book for the armchair scholar, to be perused in the comfort of one's own home and enlivened with straightforward demonstrations that can be carried out in the kitchen or garage. We have written this book because when we first started our chemistry studies, these were the types of books we loved. This book was written by chemists, but it is not written for chemists. It is written for the students we used to be and the scholars within us all. We hope the reader enjoys the following pages as much as we did the writing.

# PREFACE

> *I am beginning chemistry, a most unusual study. I've never seen anything like it before. Molecules and Atoms are the material employed, but I'll be in a position to discuss them more definitely next month.*
>
> —Jean Webster, *Daddy-Long-Legs,* 1912

Given the present state of Western society, it may be difficult to believe that sex and cooking were once subjects to be approached with trepidation. But when Irma Rombauer wrote the *The Joy of Cooking* in 1931 and Alex Comfort wrote *The Joy of Sex* in 1972, they were addressing very real societal concerns. For Rombauer it was clarification of the kitchen, and for Comfort it was the demystification of the boudoir. Nowadays, chemistry can, for some, be perceived as a likewise worrisome affair. So perhaps it is time to discover the joy of chemistry.

Within these pages, we will explore the magic inherent in chemistry—from the fascination of fall foliage and fireworks, to the functioning of smoke detectors and computers, to the fundamentals of digestion and combustion—and we'll illustrate these precepts with hands-on experiments using familiar materials. Laboratories and calculators are not required to enjoy the beauty of chemistry: the concepts can be explained in terms of everyday experience and confirmed with materials from your own closet.

Chemistry is often called the central science because it is central to our understanding of the physical and biological world and central to our common concerns, from medicine to politics to economics. Therefore, the principles of chemistry should resonate with our perceptions and experience. Chemistry can be a fascinatingly familiar science that appeals to our intuition, logic, and—if we're willing to get down and dirty—our sense of enjoyment, too. Our purpose here is to partake in the delights of chemistry and discover a little something in the process—about cars, the cosmos, and even crime.

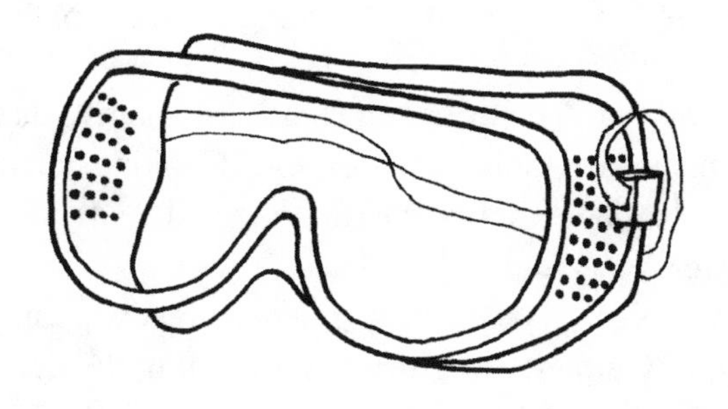

# A FEW NECESSARY WORDS ON SAFETY

*While reading a textbook on chemistry, I came upon the statement "nitric acid acts upon copper" . . . I had seen a bottle marked "nitric acid" on a table . . . I was willing to sacrifice one of the few copper cents then in my possession. . . .*

*A greenish blue liquid foamed and fumed over the cent and over the table. The air in the neighborhood of the performance became dark red. A great colored cloud arose . . . picking* (the penny) *up . . . I learned another fact—nitric acid not only acts upon copper but it acts upon fingers. . . . I drew my fingers across my trousers and another fact was discovered. Nitric acid also acts upon trousers.*

—Ira Remsen, *My First Hundred Years,* ca. 1970

The precedence for books demonstrating home chemistry has been well established, but so is the precedence for a safety discussion at the beginning.[1] True, we are all reasonable, responsible adults here, but so was the hero of the above quote. Therefore, a few necessary words on safety could not hurt. *The Joy of Cooking* offers advice on food preservation and *The Joy of Sex* offers precautions against disease, so why shouldn't chemists have their warnings, too? Cooking, chemistry, and sex share comparable dangers for the unprotected and unprepared. Let's review the basics.

If you don't own a pair of safety glasses already, go now to the hardware store and buy a pair. Wear them while doing the demonstrations in this book.

Read the safety precautions embedded in the demonstrations and take them to heart.

Resist the urge to get creative. Random mixtures of household chemicals can create some pretty nasty brews. *Never mix ammonia with bleach!*

Work away from sources of sparks, which can include, interestingly enough, cell phones or other electric appliances.

Pour waste into the toilet, not the kitchen sink or bathroom sink.

Keep the amounts to a minimum. The difference between a bang and a pop is a matter of magnitude.

Know the location of the nearest working shower, fire extinguisher, and phone, but you should know these things anyway, as a normal functioning household presents enough hazards to warrant these precautions.

Keep everything away from children. This is an *adult* chemistry set. Keep things away from pets, too.

You will be directed to use kitchen utensils in many of the demonstrations. Don't assume that this means that the chemicals are nontoxic. Use plasticware and paper plates whenever possible (you will be told when only glassware will do) and throw the paper and plastic away after use. Even inexpensive paper plates will work for many of the demonstrations if they have a thin plastic coating. If you have to use glass or ceramic utensils, wash them thoroughly and separately before any contemplated reuse with food, or, better, maintain a separate set dedicated to chemical demonstrations. You will probably do most of the demonstrations in the basement, kitchen, or bathroom, but keep the chemicals well separated from toothbrushes, foodstuffs, or counters that will come in contact with food or toothbrushes. You may cover your counters with newspaper for added protection.

Microwave ovens vary, so treat the microwave times given in the demonstrations as approximate. Get to know your own microwave oven and use mitts to remove hot items. Microwave ovens can also heat unevenly, so be aware that solutions heated in the microwave may be at the boiling point without showing bubbles. When in doubt, wait a minute after opening the door and use mitts. And, as always, be certain that you remove any metal, such as spoons, from items to be inserted in the microwave. This is chemistry, not cooking, but that does not change the physics of the microwave. Any resulting sparks may be exciting, but they are dangerous and can easily destroy a microwave.

The chemical procedures presented here have been carefully chosen to maximize your pleasure and minimize any risk, but please do not assume this is true of all chemical reactions or all chemical demonstrations. Again, experimentation is best left to the professional experimentalists and more complicated demonstrations to those trained in chemical demonstration.

Don't use any of the chemicals we discuss in a poultice or as a suppository or anything else that people were wont to do in the past or are doing in the present. Don't, under any circumstances, eat the chemicals in these demonstrations or drink them or bathe in them or splash them in your eyes. Don't expose them to flames or sparks (unless it's part of the demonstration). Protect your clothes or wear old clothes. Protect your hands. Wear rubber gloves when handling chemicals and wear yard-work gloves when handling steel wool. Treat all chemicals in this book as you would treat chlorine laundry bleach, gasoline, insecticides, and other things you know to be hazardous.

In other words, follow the same types of precautions offered in any children's chemistry set. In fact, you might go out and buy such a chemistry set. You can say it's for your children, or you can order it online. The chemistry kit will have the safety glasses you need and a few chemicals and a few experiments. It won't have the in-depth explanations, which is what we offer here, but it might be amusing and could save some trips to the hardware store.

# SHOPPING LIST AND SOLUTIONS

## SHOPPING LIST

In alphabetical order, here is a list of ingredients that will be used in the demonstrations. Most items can be purchased at a hardware store or the grocery store. If they cannot be purchased locally, they can be purchased online.

adhesive tape
alum (can be found with the spices in a grocery store)
aluminum foil
aluminum wire (can be purchased at a hardware store)
ammonia (household, clear)
analgesic stomach tablets, fizzing variety
aquarium pH-lowering solution that lists "sulfuric acid" and perhaps "water" as its only ingredients (at least three 37-milliliter bottles)
aquarium pH-testing kit (freshwater, should say "bromothymol blue" in the ingredients list)
aspirin

baking soda (*not* baking powder)
balloons
bar magnets (small)
basting syringe
black-light lightbulb and fixture
bleach (household)
candles
canola oil
chalk
citric acid (sold as "sour salt" in gourmet food shops or can be purchased online)
clear, carbonated soda (usually a lemon-lime soda will do)
clothes pins (plastic or wooden)
coat hanger (metal or plastic)
copper pot scrubber that is all copper mesh (can be found at hardware or grocery stores)
copper wire, at least two feet (60 centimeters), preferably uninsulated, but insulated can be used
corks, rubber or natural, that will fit into the end of a twenty-ounce (590-mL) plastic soda bottle
cornstarch
cottage cheese (purchase as needed)
cream of tartar (usually available on baking aisle or with spices in grocery store)
D-cell battery and battery holder (the holder is not absolutely necessary)
dropper with squeeze bulb (commonly called an eyedropper)
fingernail polish remover (the kind that contains acetone)
flexible gloves (the type used to protect the hands while washing dishes or when using household cleaners are okay, but they tend to fit poorly; surgical gloves can be purchased in a drugstore and are better)
food coloring (the liquid drops, not the squeezable gel)
galvanized nails (zinc-coated nails available in hardware stores)
garden gloves (thick cloth or leather)
gas-reducing digestive aid (the kind taken before meals to reduce intestinal gas)

glycerin (also called glycerol, from drug counter)
hydrogen peroxide
kitchen matches
lemon
liquid laundry bluing
lye (generally sold as drain cleaner, but do not get the kind that contains aluminum metal)
masking tape and permanent markers
metal spoons (two that can be discarded after use)
milk (whole milk, not skimmed or 2 percent, purchase as needed)
mineral oil
mozzarella (real and imitation, purchase as needed)
paper clips, metal
paper plates with coating
paper towels
more paper towels
plastic bags (the type that can be sealed by pressing the top together, in the sandwich size and in the gallon size)
plastic cups (clear and disposable)
plastic funnel
plastic soda bottles (clear, twenty fluid ounces [590 milliliter], at least ten)
plastic soda bottle (clear, 67.6 fluid ounces [2 liter])
plastic spoons
plastic wrap
purple cabbage (can be chopped up and frozen until needed)
rubbing alcohol
safety glasses (can be purchased from a scientific or educational supply store or hardware store)
salt petre, also called sodium nitrate (can be purchased from a cooking-supply store and some sophisticated grocery stores or online)
small plastic jars with caps (half-pint water bottles work well)
small, clear plastic containers to use as test tubes, or actual test tubes (which can be purchased at a scientific or educational supply store)
small juice glasses (three or four)

sponge (artificial or natural)
steel nails
steel wool (fine grade)
straws
super glue (the kind that contains cyanoacrylate)
swimming pool pH-testing kit (should say "phenol red" in the ingredients list)
table salt
table sugar
tall drinking glasses (three or four)
tea bag of brown tea
thermometer, range up to at least 220°F (105°C) (most meat thermometers will do)
tincture of iodine
vinegar
voltmeter (able to measure DC voltage in the one-to-three volt range)
whole milk
wooden kitchen matches

## SOLUTIONS

1. Copper sulfate

*Caution: Some fumes will be generated during this preparation, so work in an open area as you would if you were working with solvents, varnish, or paint thinner. The fumes are nitrogen-oxygen compounds such as those present in smog.*

a) Put on the safety glasses. If you have a copper-mesh pot scrubber, use this in place of the copper wire. If not, strip ten inches (30 centimeters) of copper wire if it is insulated (use as is if it is not) and coil it so it lies flat in the bottom of a plastic cup that can hold at least two cups (500 milliliters) of liquid. If you have to cut it to strip off the insulation, that is all right. The wire can be in smaller pieces that add up to about ten inches (30 centimeters). If you are using a copper pot scrubber, put the pot scrubber in the bottom of the glass. The cup can be clear plastic or clear glass, but the cup should not be reused for drinking in either case.

b) Add a third of a cup (80 milliliters) of the aquarium pH-lowering solution until the copper is well covered or emersed. This solution is acidic, so you may want to use gloves. Be careful, in any case, and do not get it on your hands or clothing. If you get it on your hands, rinse your hands thoroughly with water.

c) Add about four plastic teaspoons (about 8 grams) of salt petre (sodium nitrate) crystals. Swirl to mix. Loosely cover the cup with plastic wrap and set it aside. The solution should be ready in twelve hours or so. Although the reaction is slow, you can tell it is proceeding by the stream of small bubbles coming from the copper surface. If all has gone well, the solution will be an intense blue. There will be copper and sodium nitrate crystals left over in the bottom of the cup.

d) Using your funnel, pour off the blue liquid into a plastic bottle that can be capped. Take care not to pour any leftover crystals or copper into the bottle. Leave the leftover crystals and copper in the cup. You may want to use gloves because the solution is still acidic and you want to avoid getting it on your hands. If you do get it on your hands, rinse your hands with water.

e) Cap and label the bottle with the date, your name, and "copper sulfate in sulfuric acid." Add "Caution: Corrosive!" Store the solution at room temperature in a secure location, well away from children and pets.

f) Leave any undissolved copper and crystals in the bottom of the cup. When more copper solution is needed, you can simply replenish the ingredients in this cup. Cover the cup with plastic wrap, label it, and store it in a secure location, away from children and pets.

2. Iron acetate

a) Put on your safety glasses. Pour two cups of vinegar into a clear cup. (The cup can be plastic, but it doesn't have to be.)

Put on garden gloves (a must) and then tear off about a cubic inch (20 milliliters) of fine steel wool. This is about as much as you can tear off in one healthy pinch. Put the steel wool in the vinegar, and cover the cup with plastic wrap. Label the cup "iron acetate" and add your name and the date. The solution will be ready for use after sitting overnight. Save the solution with the steel wool still in it, as it will be used in a couple of demonstrations.

b) Store the solution at room temperature in a secure location, well away from children and pets.

3. Purple-cabbage indicator

a) Chop up about a half cup (120 milliliters) of purple cabbage, put it in a sandwich bag with a half cup (120 milliliters) of water. Heat in the microwave for approximately thirty seconds. The bag and contents should be very warm but not hot enough to scald or boil. After warming, the water should be shaded purple by the cabbage, and a bit of squeezing and manual crushing of the cabbage may make it darker.

b) The resulting solution will have a strong odor, which will get stronger on aging. To avoid overly unpleasant smells, the indicator can be made up only as needed, frozen if made up ahead of time, or can be made from dark purple plums if these are available.

But chemistry can be a smelly business!

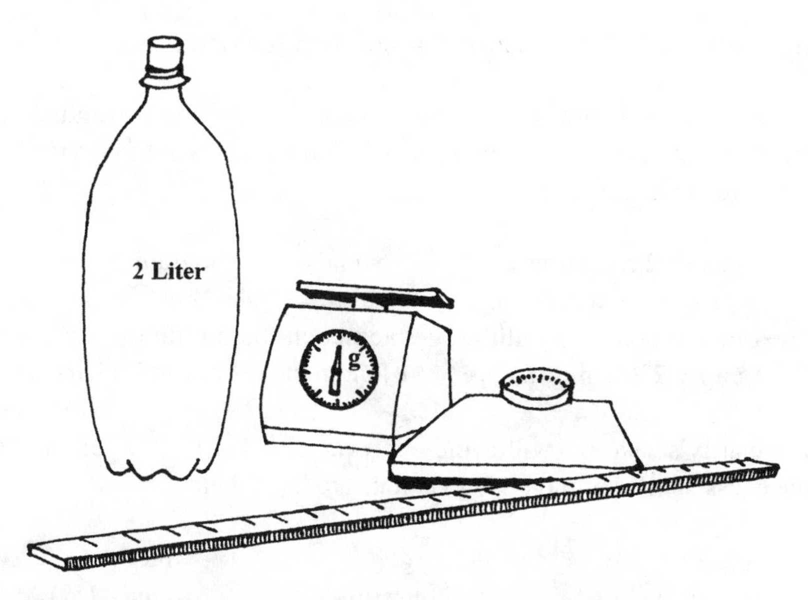


# A MOMENT FOR METRIC

Metric units (grams, centimeters, milliliters), as opposed to traditional US units (ounces, inches, cups), are preferred for scientific measurements. There is a valiant effort in the United States to move toward the use of metric units in everyday applications; however, most people in the United States still use traditional units for common purposes. Throughout this book, US units will be used, followed by their approximate metric equivalent in parentheses. Though no doubt metric units are now second nature to many, they are foreign to as many more, so a set of comparisons is offered here. Bear in mind that a centimeter is about the width of a small paper clip, a penny weighs about two grams, a meter is a little longer than a yard, and a milliliter is (very roughly) a quarter teaspoon. You may also want to keep in mind that an ounce is a measure of mass, but a fluid ounce is a measure of volume—which reminds us of the complexity of US units. So even if metric is not second nature now, don't worry. Anyone who has mastered the subtleties of US units shouldn't have much trouble with metric.

Some useful approximate US/metric conversions are given below. In the demonstrations, approximate metric equivalents will be provided with measurements in US units.

| | |
|---|---|
| 1 inch = 2.5 centimeters | 1 pound = 0.5 kilogram |
| 1 quart = 0.9 liter | 1 ounce = 28.4 grams |
| 1 tablespoon = 15 milliliters | 1 teaspoon = 5 milliliters |
| 1 cup = 236.7 milliliters | 1 fluid ounce = 29.5 milliliters |

For clarity's sake we will avoid abbreviations, but for the sake of completeness some common abbreviations are listed below.

| | | |
|---|---|---|
| cup (c) | gram (g) | liter (L) |
| meter (m) | milliliter (mL) | ounce (oz) |
| pound (lb) | tablespoon (tbsp) | teaspoon (tsp) |

Scientists also normally use the Celsius temperature scale, but we will report temperatures in both degrees Celsius and the traditional US Fahrenheit temperature scale. For convenience, a comparison of some common temperatures is given below.

| | room temperature | boiling water | frozen water | normal body temperature |
|---|---|---|---|---|
| Fahrenheit | 68° | 212° | 32° | 98.6° |
| Celsius | 20° | 100° | 0° | 37° |

## Introductory Demonstration: Bottle Rocket and Oobleck

*Their faces were as rosy as if they had been exposed to the roaring flames of an oven; their voices resounded in loud accents; their words escaped like a champagne cork driven out by carbonic acid.*

—Jules Verne, *Around the Moon*, ca. 1870

Let's begin with a bang. A bottle rocket.

Put on your safety glasses. Tear a paper towel into a strip about four inches wide and seven to eight inches long. Take a rounded plastic spoonful of baking soda (about 2 teaspoons or 10 milliliters) and spread it down the length of the paper-towel strip, in the center. Leave an inch or two at the end of the paper towel free of baking soda. Roll the paper towel lengthwise, as if rolling a cigarette, making sure that the baking soda doesn't get close to the ends of the towel. The roll should be small enough to go through the end of a plastic soda bottle. Once it is rolled, twist the ends so that the powder does not come out. When you are done, your paper towel roll should look like a cigar with twisted ends.

Remove the label from a twenty-fluid-ounce (590-milliliter) clear plastic soda bottle, rinse it out, and shake out any excess water. Find a cork that fits snuggly into the bottle opening and make certain it seals the bottle well without needing to jam it in. As extra insurance, pull a balloon over the cork (put the cork *inside* the balloon) so that you have a rubber-coated cork that will make a nice tight seal with the top of the soda bottle.

Pour a half cup (120 milliliters) of vinegar and a half cup (120 milliliters) of water into the bottle and swirl to mix. A two-liter soda bottle may be substituted if you use twice as much vinegar and water. Add enough purple-cabbage indicator so that the resulting solution is bright pink, but do not add more than one tablespoon.

Find a location, preferably outside, where it will not matter if a cork shoots upward with respectable force. Put the soda bottle with vinegar on the selected spot, drop in the twisted tube of baking soda, and rapidly and firmly cork the bottle. Stand the bottle upright so the cork will fire straight up in the air and out of the way of children, pets, or neighbors. Stand back. There should be a few second's delay while the vinegar solution penetrates the paper towel, then there should be a vigorous foaming and a satisfying bang as the cork shoots out. At this point the solution in the bottle will have changed from pink to blue or violet (depending on the amount of baking soda used). If the reaction appears to die down after a minute instead of popping the cork, you can swirl the bottle or you can push at the bottom of the cork to dislodge it as it was probably jammed too tightly into the opening. It may take a few attempts to get it right, but it's worth it.

The reaction that pops the cork is

$$\text{baking soda} + \text{vinegar} \rightarrow \text{carbon dioxide} + \text{water}$$

Carbon dioxide is a gas and tends to occupy a larger space than the solution. The expanding gas builds up a force sufficient to expel the cork from the mouth of the bottle.

Gas production is a good indication that a chemical reaction has taken place. Color change is also an indication of a chemical reaction; however, in this case the reaction could have proceeded just as well without the purple-cabbage indicator and without changing color. The purple-cabbage indicator is added to show that the reaction is an acid-base reaction, which we will discuss shortly.

The defining feature of a chemical reaction is that you start with one type of material or materials (a sour, pink liquid and a salty solid in this case) and end up with another type of material (a violet, watery liquid and a gas). There are other signs that can indicate a chemical reaction has occurred, such as a change in temperature, the production of light, or the formation of a solid that settles out of a solution—and in future demonstrations we will see all these and more. But to understand how the processes occur, we need a solid background in the basic structure of matter, and that is where we begin.

That is, after one more demonstration and this time with a splat—from oobleck.

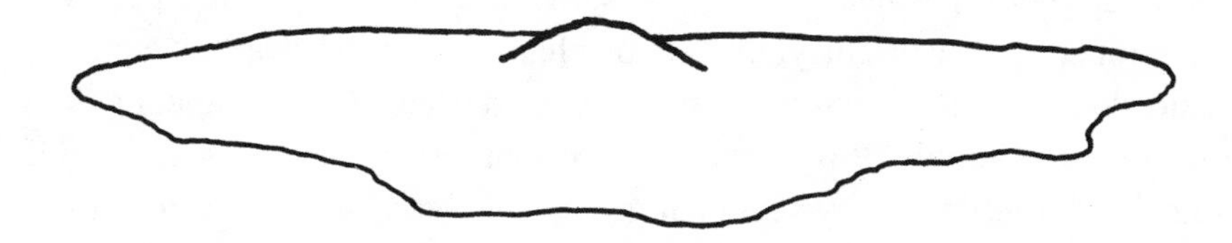

Start with a cup (240 milliliters) of cornstarch in a small bowl and add a third cup (80 milliliters) water. Using a metal spoon (a plastic spoon might break), stir the water into the cornstarch. The mixture will quickly become thick and difficult to stir, but persevere. The exact amounts of water and cornstarch needed will depend on the brand of cornstarch, so you may have to add a little water or sprinkle in a little more cornstarch. You want to end up with something approximately the consistency of wet cement or corn-bread batter, but slightly dryer. Cutting through the mixture with the spoon helps the mixing process. When you are done, you have what science teachers have come to call *oobleck*, after the Dr. Seuss book *Bartholomew and the Oobleck.*[1]

Try smacking the spoon down on the oobleck, and then try picking up the oobleck in your hands. It feels fairly solid when you pick it up and squeeze it, but if you hold it in the palm of your hand, it appears to "melt" into a puddle and you can drip it between your fingers. Oobleck behaves in this manner—not quite solid, not quite liquid—because cornstarch traps water, somewhat like a sponge, but the structure of starch is less rigid so it can flow with its water load.

Try rolling up a pinch of the oobleck into a ball and throwing it on the floor. It should bounce a bit then hit the floor with a splat and puddle. Try slicing through the oobleck with a knife. It cuts somewhat like cheese, appearing more solid than liquid. Try pulling the oobleck apart quickly and then slowly. When the oobleck is separated quickly, it behaves more like a solid; when it is pulled apart slowly, it behaves more like a liquid.

Although the properties of oobleck seem quite different from the cornstarch and the water from which it was formed, the formation of oobleck is not a chemical reaction; it is a physical change. The cornstarch has trapped the water, but the water can be removed by the simple process of letting the oobleck dry out (though it may take several hours). The *physical* properties of the cornstarch and water have changed, but the *chemical* properties have not.

Changing a physical property is something like changing a person's appearance. You can cut your hair or change your clothes, but you are basically the same person as when you started. The change can be reversed by simply letting your hair grow out or by changing back into your former attire. A physical change for a material such as water might be to boil or freeze it or divide it into separate cups. Chemical changes, however, are a fundamental change of one type of material into another. A chemical change results in a new material with new physical and chemical properties. The bottle rocket represents a chemical change because the vinegar and baking soda turned into new materials: carbon dioxide and water. No amount of drying will recover the baking soda.

Because most chemical reactions also involve physical change, we will be discussing both, but our emphasis will be on the chemistry. As chemists, of course, we like a good splat, but nothing beats a bang!

# INTRODUCTION

## It Begins with a Bang . . . and a Splat

> *It's chemistry, brother, chemistry! There's no help for it, your reverence, you must make way for chemistry.*
>
> —Fyodor Dostoevsky, *Brothers Karamazov*, 1880

> *I must go to the laboratory and look into a little matter of acids and salts and alkalis. I've burned a hole as big as a plate in the front of my chemistry apron, with hydrochloric acid. If the theory worked, I ought to be able to neutralize that hole with good strong ammonia, oughtn't I?*
>
> —Jean Webster, *Daddy-Long-Legs*, 1912

We lead off with these quotes because they illustrate two observations. First, if one judges by the number of references to chemistry in Western literature of the late nineteenth and early twentieth centuries, it would appear that chemical literacy was once more common. From Arthur Conan Doyle to Emily Dickinson, many of these earlier authors displayed a friendly acquaintance with the principles of chemistry and assumed a familiarity on the part of their readers, too. Such allusions to chemistry, unfortunately, seem to be less frequent in contemporary literature, presumably because of a lack of chemical fluency on the part of the authors and the readership. It is our belief that this situation might be alleviated by providing more in the way of public access to chemistry, which is the objective of this book.

The second observation illustrated by these two particular quotes is that some people, past and present, are naturally drawn to chemistry—and some are not. Why the difference? Is there a variation in brain cells? Is there an undiscovered chemistry gene? Maybe. But probably not. A better explanation may be found in the following abbreviated biographies of chemistry Nobel Prize winners and obvious aficionados of chemistry.

Linus Pauling (1901–1994), son of an Oregon pharmacist, shared a friend's chemistry set when he was fourteen years old.[2]

Mario Molina (1943– ), who worried about atmospheric chemistry as his work carried him over the globe, had an aunt who helped him set up a laboratory in the bathroom of his boyhood home in Mexico City.[3]

Dorothy Hodgkin (1910–1994), a British chemist with an archaeologist father and a botanist mother, grew crystals with a soil-chemist test kit when she was ten years old.[4]

As a young Texan, Robert Curl (1933– ) used up the chemicals in the chemistry set his parents bought him for Christmas, so he found a sympathetic pharmacist to help him resupply. His parents rethought the wisdom of the gift when, in using his new stock, he permanently altered the porcelain top on his mother's stove.[5]

Not much in the way of a shared gene pool, but perhaps some other traits in common. For one, these successful chemists (as well as many others) seem to have first encountered chemistry as an amusement. Does this mean that to appreciate chemistry as an adult you should choose parents who give you a chemistry set as a child? No. It is never too late to indulge yourself in the joy of chemistry. Each of our budding Nobel chemists also had a strong curiosity, and if you have picked up this book, then you do, too—and it is our hope to gratify this curiosity. We are proud to present herein a virtual adult chemistry set complete with demonstrations. The critical difference with our approach and that of most chemistry sets, however, is that the results are explained and the demonstrations are integrated into an account of the science of chemistry. A description of the expected outcome will be given with each of the demonstrations, as well as an explanation of the principles being illustrated, so the reader may, in fact, choose to enjoy this book without performing all, or any, of the procedures given in the demonstrations. But the temptation to join in the fun—with a bang and a splat—may be difficult to resist.

The above short bios might also imply that the successful chemist must have access, as a child, to a mentor or companion with whom to discuss the chemical arts. Does this mean you have to maintain a discussion group to appreciate the principles of chemistry? Not necessarily. We have provided a sort of virtual companion in the "For Example" sections, which present topical essays germane to the chemical theory being discussed. We include these to help the reader make intuitive connections with the concepts of chemistry and to dispel the notion that chemistry is somehow removed from everyday experience.

Take, for instance, our mundane experience of driving cars. What does chemistry have to do with cars? As it turns out, quite a lot.

## INTRODUCTORY EXAMPLE: CHEMISTRY AND CARS

Throughout this book, we will discuss chemistry in situations as diverse as cooking and solving crime. By way of introduction, our first "For Example" is going to take a quick walk through many of the topics that will be covered in these pages and illustrate them using an icon of modern technology: the automobile.

The *Joy of Chemistry* consists of two parts: Part 1 presents the fundamental principles of chemistry, and part 2 is a survey of some of the specialized fields of chemistry in which these principles are applied. Chapters 1 and 2 of part 1 introduce atomic structure with the chemist's best friend: the periodic table. Atomic structure, as reflected by the periodic table, determines the makeup of chemical compounds—such as those that supply the structure and energy for the petroleum-based internal combustion engine. The type of reaction that is responsible for the power of this engine is the same type of reaction that supplies metal for its construction—oxidation-reduction reactions—which we'll describe in chapter 3. The connection of chemistry with automobile exhaust is made in chapter 4, which examines acid-base chemistry. The chemistry that clogs up the radiator is touched on in chapter 5, in the discussion of precipitation reactions. The properties of the many materials that make up an automobile are a consequence of chemical bonding, discussed in chapter 6. The control and predictability of the various chemical reactions that take place in a car are governed by the principles of chemical reactions, the topic of chapter 7. The properties of automotive lubricants are explained in terms of intermolecular forces, which are the subject of chapter 8. The topic of chapter 9 is concentration, and a car's choke regulates the concentration of oxygen in the air/fuel mixture for optimum combustion. The explosive reaction that changes gasoline to gas—and the gas properties that push the piston—are topics for chapters 10 and 11, which describe the gas phase of matter and reactions involving this phase.

Soaps and surfactants, useful for washing and waxing cars, come within the scope of chapter 12, which describes solid-state and surface chemistry. Chapter 13, on thermodynamics, delves into the operation of car-powered refrigerators. Metal fatigue is a topic for chapter 14, which investigates phase changes for pure materials and mixtures. And after reading chapter 15, a discussion of chemical equilibrium, the reader will be better positioned to understand why virtually all the gasoline turns to gas. The colligative properties of solutions, explored in chapter 16, can be applied to the salting of icy roads and the use of antifreeze in the radiator. The treatment of chemical kinetics in chapter 17 will relate to engine knock and catalytic converters. The discussions of photochemistry and electrochemistry in chapter 18 will shed light on automobile paint and chrome plating, the finishing touch.

In part 2 we will take a look at the specialized areas of chemistry and see that here, too, are intimate connections to daily living—and daily driving. The organic chemists of chapter 1 of part 2, who have turned the rich petroleum stew into plastics and polymers for interiors, gadgets, hoses, and tires, are today facing a new challenge: the development of comparable materials from a new feedstock, either the biomass or recyclables—an exciting and interesting challenge for the creative chemical scientist. As we will see in chapter 2 of part 2, the inorganic chemists have more of the periodic table to choose from and have every bit as much challenge. Consideration of their work will lead us to a discussion of how fuel cells may help us in the future—and how radiator cleaners work right now.

The concerns of the biochemist, as outlined in chapter 3 of part 2, address the automobile's most dangerous component—the one located directly behind the wheel. But biochemists are also concerned with the sources and management of alternative materials harvested from the biosphere. In chapter 4 of part 2, we consider the efforts of analytical chemists and find that these intrepid souls draw from all of the above described disciplines. It is the job of the analytical chemist to certify the consistency and quality of many materials, including materials for cars, and in some instances, the consistency and quality of human life—as we will see when we consider the contributions of this field to crime-solving as well as cars. In chapter 5, we peer into our crystal ball to see the future of chemistry: a universe of new sources and substances. These new materials will no doubt lead to remarkable innovations—and, of course, new cars!

Ready? Let's open the hood and take a look. . . .

# PART I

# INTRODUCTION

## Theory, Octaves, and Scales

Not unlike music and literature, chemistry is described in terms of its elements and has a theory based on fundamental principles. And, as with music and literature, there is much in chemistry that is art as well. Looking at nature is like looking in a mirror, and we interpret what we see in light of our experience and from our own perspective. The principles of chemistry represent an attempt to draw from nature a system and, as such, must remain a bit blurry about the edges and able to accommodate exceptions to the rule. Yet, over the years, there has evolved a substantial repertoire of physical models and methods that allows us to describe, understand, and predict the behavior of a considerable body of materials with a respectable range of properties. So we'll begin with the basics of this theory. For instance, we all know that chemists use scales, but did you know that chemists think in octaves, too? In part 1 of *Joy of Chemistry*, we'll learn why.

## Demonstration 1: Water Witch

*For days he made ceaseless calculations, but they were all wondrously unsatisfactory. . . . He finally concluded that the only way to prove himself was to go into the blaze . . . to watch his legs to discover their merits and faults. . . . To gain it, he must have blaze, blood, and danger, even as a chemist requires this, that, and the other.*

—Stephen Crane, *Red Badge of Courage*, ca. 1895

Take a plastic spoon and rub it in your hair or on a sweater until the spoon acquires a static charge, as evidenced by the attraction of the spoon for the hair or fibers on the sweater. Turn on a faucet so that there is a very thin stream of water. Hold the spoon close to the water and you will see the stream of water bend. What happened? Electron transfer.

Friction will cause electrons to transfer from one material to another if one of the materials has a stronger attraction for electrons. Most plastics have a stronger attraction for electrons than hair and clothing, so the direction of transfer was probably from your hair or sweater to the spoon, allowing the spoon to acquire a negative charge.[1] It could be a positive charge if the transfer were in the other direction, but the effect on the stream of water would be the same. Water is made up of atoms of hydrogen and oxygen, which have negative electrons in clouds surrounding positively charged nuclei. When the electric field of these charged particles interacts with the static electricity on the plastic spoon, the stream of water is drawn toward the spoon.

Find this idea of electrons bewitching? Read on.

# CHAPTER 1

## Electrons and Atoms, Elephants and Fleas

*Our science is sensual. . . .*

—Ralph Waldo Emerson, ca. 1830

In the above epigraph, Emerson was not intimating that science is sexy; he meant that we base our scientific theory on the input of our senses: touch, taste, hearing, sight, and smell. But when it comes to the theory of atoms, our senses fail. We can't see an individual atom, we can't taste one, we can't feel one, we can't hear one, and we can't smell just one. If someone hits us on the head with a single atom of tungsten (fairly hefty as far as atoms are concerned), we remain blissfully unaware.

The reason we are so insensitive to single atoms is that a single atom is extremely small. Some ten million individual atoms would have to line up to span the length of a grain of rice. The parts that compose atoms are smaller yet. The nucleus, or center of an atom, is made up of protons and neutrons, and the radius of a proton is on the order of a femtometer, which is a quadrillionth of a meter, or a millionth of a billionth of a meter. Mighty small. The protons and neutrons have a mass of about a septillionth of a gram (a trillionth of a trillionth), which is diminutive in anybody's book. The electrons are about two thousand times less massive than the protons and neutrons. Electrons are to protons as fleas are to an elephant—the proton being the elephant. When we calculate the mass of an elephant, we don't add on the mass of the elephant's fleas, and when we calculate the mass of an atom, we don't add the mass of the electrons.

So the question naturally arises: If atomic particles are so small we can't use our senses to detect them, how do we know they are there? By inference. Humankind learned long ago that the input of the senses can be flawed. Optical, olfactory, and tactile illusions abound. So to discover the nature of those parts of the world that cannot be smelled, touched, and seen, people have learned to look at secondary effects and infer their causes. The concept was well captured by the venerated scientist Ernest Rutherford in the advice he gave to James Chadwick when Chadwick was looking for evidence for the neutron. Rutherford advised,

> How could you find the Invisible Man in Piccadilly Circus? . . . [B]y the reactions of those he pushed aside.[2]

Similarly, J. J. Thomson, making inferences from his own work and the work of others, declared the existence of electrons in 1897. By consensus, electrons were assigned a negative charge. Thomson was able to determine the amount of charge on a given mass of electrons by bending a beam of electrons in a magnetic field.[3]

This interaction of electrons with a magnetic field could cause a television picture to distort in the presence of a magnet. The electron beam in the cathode ray tube (CRT or television tube), which causes the phosphorescence on the screen, bends in a magnetic field. Of course, one should not bring a magnet up to a TV screen unless one has a dispensable TV because the interaction could do permanent damage to the electronics. But if one has a dispensable TV, it is an interesting effect to witness.

However, J. J. Thomson did not irrefutably establish the particulate nature of matter. It remained until 1909 for Jean Perrin to provide the definitive evidence for atoms, which he did by measuring the motion of microscopic pollen particles suspended in water. His detailed observations of this Brownian motion (named after the botanist Robert Brown) could be explained if it were assumed they were being buffeted about by moving atoms. His observations convinced the scientific community of the validity of the atomic model. Of course, they had been using the model successfully before Perrin, but it was nice to have such elegant confirmation.[4]

In 1910 Ernest Rutherford realized that atoms must be composed of a central, dense nucleus surrounded by a lot of empty space. He fired some atomic-sized particles at an ultrathin gold foil and found that most of the particles passed through the foil, but a few bounced back. Always able to turn an interesting phrase, Rutherford commented, "It was . . . as if you had fired a 15-inch shell at a piece of tissue paper and it came back and hit you."[5]

Rutherford suggested that this nucleus at the center of the atom was composed of densely packed positively charged particles. Soon after, Henry Moseley, before his early death at Gallipoli in World War I, supplied experimental evidence for these particles, the protons.[6] The other particles in the nucleus, the neutrons, proved a bit harder to pin down because they have no charge. But James Chadwick, taking Rutherford's advice, finally confirmed their existence in 1932. Chadwick measured the rebound of certain radiation from nitrogen and helium and found it corresponded to a neutral particle with about the same mass as a proton.[7]

It took until the 1930s to discover all the pieces of the atom, which shows how difficult it was. Even so, laying out all the pieces still didn't solve the puzzle. There are other features of the atom—besides its size—that boggle the imagination. For one, there is the density of the nucleus. The density of a substance is the mass of a given volume. For instance, a bushel of feathers and a bushel of pebbles have the same volume but a definitely different mass. The bushel of pebbles is heavier; therefore, the bushel of pebbles is denser. We mentioned that Rutherford found that atoms have densely packed nuclei, but we didn't say how dense. It turns out that a nucleus, for all its diminutive proportions, packs some whopping million trillion grams per cubic centimeter. Compare that, say, to the density of lead, which is about eleven grams per cubic centimeter. The difference is that a nucleus consists of closely packed particles, but an atom is mostly empty space. A comparison can be made with Earth and the sun. If the nucleus of a lead atom were Earth, then the distance to the nearest electron would be about the distance to the sun. That is to say, most of the atom is just empty space. If you were to pack just nuclei into a container, you would be packing in solid marbles of matter. When you pack lead into a container, you are, in essence, packing in bubbles of empty space.

The question that may immediately spring to mind is this: If matter is composed of tiny bubbles, then why don't we collapse into puddles? An unsatisfying answer is that the electrons stay where they are because they are in orbit around the nucleus. But, one might object, if the electrons have a negative charge and the protons have a positive charge, shouldn't these opposite charges attract and shouldn't the electrons come crashing into the proton?

As it turns out, this is no trivial question.

In the early 1900s, the noted physicist Niels Bohr helped to answer this question by showing that an electron could remain at a distance from a proton as long as it kept moving. An analogy could again be made with Earth and the sun. Gravity does pull these two bodies together, but Earth, by moving in its orbit, can keep falling toward the sun but never fall into it. A similar picture can be painted for the electron. The electron can be thought of as circling the nucleus and thereby being pulled toward the nucleus, but not into it.

The reader has no doubt noticed that the above paragraphs contain quite a few qualifying phrases, such as "Niels Bohr . . . helped to explain" instead of "Niels Bohr explained" and "an electron can be thought of" rather than "an electron is." This approach is necessary because the analogies are just that—analogies—and they work only at a rather crude level. The analogies fail, and fail quickly, when any degree of precision is required. We can have no exact analogy because the physics at the atomic level is quite different from the physics we experience in our relatively gargantuan, everyday, macroscopic world.

So it is with this disclaimer in mind that we present the structure of the atom as it is currently understood. We first note the main features that are of common knowledge. The nucleus, or center of the atom, is composed of positively charged particles called protons and uncharged, neutral particles called neutrons. Electrons, as is customarily acknowledged, are the negatively charged particles that reside in "orbits" around the nucleus.

Here the term *orbit* is being used metaphorically, not literally. Though the common picture is to show electrons orbiting, satellite-like, around the nucleus, the space occupied by the electrons cannot be clearly delineated. The best we can do is to describe a sort of fuzzy region of probability in which we believe the electron may be found. To emphasize this difference, we call the space occupied by electrons around a nucleus an *orbital*, not an orbit. These orbitals can be spherical, a sort of dumbbell shape, or intricate structures of loops, doughnuts, and lobes. Once again, however, there is nothing in common experience that is really quite like them, so the analogies are not perfect.

The problem only becomes worse when you include more than one electron in the discussion, which is true for every element beyond hydrogen. Electrons are charged particles, and charged particles tend to be attracted to each other if they are of opposite charge and repelled from one another if they are of the same charge. A fairly good model for this behavior is found in the behavior of magnets. Like poles of magnets repel, while opposite poles attract. In an atom, the situation is much murkier because there are not just magnetlike interactions of one electron to one electron, or one electron to one proton, but a sea of negative and positive charges that are all interacting. The Nobel physicist Enrico Fermi once likened the situation to boats bobbing in a harbor. We know intuitively that the motion of one boat will influence the motion of all the others, and vice versa, but in ways so intertwined and convoluted that the final motion of just one boat cannot be predicted at any one time.

This problem is called the *three-body problem* by the people who study such things (the theoreticians of quantum mechanics). When you have two particles in motion that attract each other, you can describe the situation with an equation. But when you have three particles, and there are attractions and repulsions, and all these particles are in motion, there are too many things going on for one neat equation. The problem is one of clouds: a cloud exists and we can point to it and measure it, but to predict in advance just where it will be and what form it will take is not possible. There are too many factors, too many variables, many of which are unknown or unknowable. This problem is at the heart of the probability approach to atomic structure.

But luckily you don't have to know the position of every cloud to predict the weather.[8] Based on the probability approach, the theory of quantum mechanics is able to accurately interpret and predict many of the properties of atoms and molecules and how they interact. Scientists have also been able to understand and work with another interesting beastie: the *ion*. An ion is an atom or molecule that has lost some electrons or gained some extra electrons, as our plastic spoon gained some electrons in the "Water Witch" demonstration. The fact that there are too few or too many electrons means that the positive charge from the protons is unbalanced and the ion has a net positive or negative charge. In the case of the plastic spoon, electrons were moved by friction, and the spoon acquired a net negative charge. Other charged species are able to perform other wonderful tricks. For instance, the miracle of photocopy machines.

## For Example: Protons and Photocopiers

Okay, some people might consider it an exaggeration to call a photocopy machine a miracle, but for anyone who has dealt with mimeograph machines, miracle is almost too weak. Photocopy machines are nearly as integral to the information age as computers and satellite cell phones. For all the photocopiers' modern-age convenience and programming—as well as features that allow one to copy, collate, stack, staple, and punch holes—the technology underlying them is actually fairly straightforward. The basic principle behind the operation of photocopiers is that static charges attract.

The parts of a photocopier are familiar to everyone who has ever used one to any extent because anyone who has ever used one to any extent has had a paper jam and had to open the copier to remove the jammed paper. If you have somehow avoided this experience, then simply go to a machine and open it to familiarize yourself with the interior workings. Integral to the photocopier is a rotating drum, a movable light source, a black powdery mess called toner, a heat source, and an elaborate system of rollers.

The first step in the photocopy process is for the drum to acquire a uniform static electric charge. The method used is slightly fancier than the method by which the spoon acquired its static charge in the "Water Witch" demonstration, but just slightly. With this static charge, the drum can now attract toner, just as the spoon in the "Water Witch" attracted water. If this were the last step, however, then the paper produced would be a uniform black color, which is not the promised copy. To accomplish the copy, a very bright light passes under the paper to be copied.

Light is used because, as we will see, light and matter routinely interact: camera film reacts with light to form an image; digital sensors record light levels in a digital camera; photosensors detect the presence of a person in a light beam and trigger a door to open. When the light strikes a dark part of the paper, it is absorbed, but when the light strikes a white part of the paper, the light is reflected onto the drum. The interior material of the drum is photoconductive, which means light will cause the interior to eject an electron. Wherever an electron is ejected from the interior of the drum, it neutralizes the static charge at the surface of the drum in just that one spot. The rotation of the drum is synchronized with the movement of the light under the paper so the flat image can be transferred to the curved surface of the drum.

The light-exposed drum then turns past the toner, and toner is attracted to the still-charged portions of the drum. A piece of paper with a static charge is now passed over the surface of the drum and attracts the toner away from the drum. The paper is heated to fix the toner onto the paper, and the new copy is presented.

If the principle behind the operation of photocopiers—the attraction of materials to a static charge—is so basic and well understood, why did it take so long to produce a practical photocopier? The answer, as with many such innovations, is that the fundamental concept existed long before the materials necessary to implement the idea. Materials science is a discipline all to its own because of the virtually infinite variety of properties that the elements, as well as the substances derived from the elements, can display. Consider, for instance, the variation in the behavior of aluminum and copper (both of which are used in electrical wiring) and steel nails and galvanized nails (both of which are hit on the head). In the next chapter, we lay the differences on the table—the periodic table, that is.

## Demonstration 2: Coppers and Robbers

*You're capable of anything, you and Osmond. I don't mean Osmond by himself, and I don't mean you by yourself. But together . . . like some chemical combination.*

—Henry James, *Portrait of a Lady*, ca. 1880

*Be careful with this demonstration. The solution you will be making is caustic and can damage clothing, surfaces, eyes, and skin. If any is spilled, rinse the area immediately with plenty of water. If any should get in your eyes, rinse your eyes with copious amounts of water and seek medical attention. For this demonstration, it is imperative to have proper eye protection.*

Put on your safety glasses and protective gloves. For the first part of this demonstration, take two clear plastic cups and add tap water to a depth of one inch (2.5 centimeters). Cut one-inch (2.5-centimeter) strands from the aluminum wire and the copper wire. Remove the insulation on the wires, if there is any. Place the aluminum wire in one cup and the copper wire in the other. Using a plastic spoon, add a half teaspoon (2 milliliters) of the lye specified in "Shopping List and Solutions"—the crystal drain cleaner without added aluminum metal. Swirl the cups gently to mix. There will be some small bubbles as the lye dissolves.

After about fifteen or twenty seconds, small streams of tiny bubbles should be seen coming from the aluminum wire but not from the copper wire.

Copper and aluminum wire seem similar. Both are used in electrical wiring. But the aluminum visibly dissolves in this caustic solution, and the copper wire does not. The bubbles you observed rising from the dissolving aluminum are hydrogen, which explains why thin strips of aluminum foil are added to some lye-based drain cleaners: to provide agitation to help break up clogs.

Keep your safety glasses on for the second part of this demonstration. Take the copper sulfate solution prepared as outlined in "Shopping List and Solutions" and pour it into a plastic cup to a quarter-inch (0.5-centimeter) or less depth. Place a steel nail and a galvanized nail in the cup, taking care that they do not touch. Both the steel nail and the galvanized nail acquire a copper coating in the copper solution, but the steel nail has a more rapid reaction. Moreover, its coating of copper will look more like copper than the coating on the galvanized nail. Steel nails are mostly iron, and galvanized nails are steel nails that have a corrosion-resisting zinc coating. The reactions these nails underwent are called displacement reactions because one metal (zinc) is displacing another metal (copper) from solution.

Both the nails perform the physical function of fastening boards, but their chemical makeup, as determined by the positions of iron and zinc on the periodic table, accounts for the difference in chemical behaviors seen here.

# CHAPTER 2

## Periodically Speaking

*That army could not recover anywhere. Since the battle of Borodino and the pillage of Moscow it had borne within itself, as it were, the chemical elements of dissolution.*

—Leo Tolstoy, *War and Peace*, ca. 1866

When taking a picture at a family reunion, a photographer may ask members of immediate families to stand together and may arrange them by height or generation. A fastidious photographer might also position the residents of the West Coast on one side and the East Coast on the other and have the Northerners stand a bit to the back and dwellers of the South in front. Such arrangements would take some time and effort but would also ensure that future generations could discern a good deal of information about any one particular subject by finding his or her position in the photo. The periodic table is similar to such a family-reunion photo: once the principle of the arrangement is understood, information about a particular element can be obtained by simply knowing its position in the table.

The clever photographer who arranged the subjects for the periodic table was Dmitri Mendeleev. Born in Serbia in 1834, he would have understood the need for organization at a family reunion, as he was the youngest of fourteen children. He also would have understood the benefit of organization by region because his mother, seeing his talent, used her limited resources to relocate him to St. Petersburg for his education. After securing a teaching position, following the manner of teachers of all time, Mendeleev began searching for a way to organize his material. His subject was the chemical behavior of the elements, and the pedagogical device he came up with grew into the periodic table.[1]

To organize the elements, Mendeleev wrote down all their observed behaviors on cards and arranged the cards into groups with similar behavior. Then he added a second level of organization and began arranging the cards according to their characteristic mass.

The terms *mass* and *weight* are often used interchangeably, but there is a subtle yet important difference. The weight of an object is a measure of how strongly gravity pulls on the object. Because the gravity of the moon is less than the gravity of Earth, objects weigh less on the moon. Mass, however, is measured on a balance, relative to some standard. The statue of "blind justice," used to symbolize the rule of law, is a blindfolded woman holding such a balance. To measure the mass of an object, the object is placed on one side of the balance, and standards of known mass are placed on the other side until the sides are level. Because gravity affects the standards the same way it affects the object being measured, the measured mass would be the same on Earth as on the moon.

By the time of Mendeleev it was known that elements combined with each other in set ratios, the way ingredients are combined in a recipe. So just as a recipe for cake might call for one egg and one cup milk, a recipe for table salt might call for one part sodium and one part chlorine. The mass of "one part sodium" was taken as the characteristic mass of sodium, and it was this characteristic mass for each element that Mendeleev used for his second layer of organization.

Other investigators had arranged the elements in groups according to their characteristic mass, but Mendeleev took the bold step of keeping groups together that had similar chemical behavior, even when it meant leaving gaps in the table. The genius of Mendeleev's insight was quickly revealed when others found the missing elements and fit them into the blanks in his table.

Several decades later, when other investigators deciphered the structure of the atom, they found that trends in atomic structure followed the arrangement in the periodic table, revealing the connection between atomic structure and chemical behavior. The concept may seem obvious in hindsight: the behavior of everything from fish to flashlights is determined by how they are put together. But the underlying structure of the atoms—electrons, protons, neutrons—was unknown when Mendeleev constructed his table.

The modern periodic table, considerably expanded since Mendeleev's time, is shown in figure 1.2.1. Many of the symbols may be familiar, such as H for hydrogen and O for oxygen, but others may not. Some elements such as tungsten (W) and sodium (Na) derive their symbols from alternate names for these elements. Tungsten's *W* is from the German *Wolfram.* Sodium's *Na* is from the Latin *natrium.* The numbers above the symbols are the atomic numbers (a concept to be elaborated upon shortly). Listings of the elements by name, by symbol, and by atomic number are given in the appendix.

| 1 H | | | | | | | | | | | | | | | | | | | | | | | | | | | | | | | 2 He |
|---|---|---|---|---|---|---|---|---|---|---|---|---|---|---|---|---|---|---|---|---|---|---|---|---|---|---|---|---|---|---|---|
| 3 Li | 4 Be | | | | | | | | | | | | | | | | | | | | | | | | | 5 B | 6 C | 7 N | 8 O | 9 F | 10 Ne |
| 11 Na | 12 Mg | | | | | | | | | | | | | | | | | | | | | | | | | 13 Al | 14 Si | 15 P | 16 S | 17 Cl | 18 Ar |
| 19 K | 20 Ca | 21 Sc | | | | | | | | | | | | | | | 22 Ti | 23 V | 24 Cr | 25 Mn | 26 Fe | 27 Co | 28 Ni | 29 Cu | 30 Zn | 31 Ga | 32 Ge | 33 As | 34 Se | 35 Br | 36 Kr |
| 37 Rb | 38 Sr | 39 Y | | | | | | | | | | | | | | | 40 Zr | 41 Nb | 42 Mo | 43 Tc | 44 Ru | 45 Rh | 46 Pd | 47 Ag | 48 Cd | 49 In | 50 Sn | 51 Sb | 52 Te | 53 I | 54 Xe |
| 55 Cs | 56 Ba | 57 La | 58 Ce | 59 Pr | 60 Nd | 61 Pm | 62 Sm | 63 Eu | 64 Gd | 65 Tb | 66 Dy | 67 Ho | 68 Er | 69 Tm | 70 Yb | 71 Lu | 72 Hf | 73 Ta | 74 W | 75 Re | 76 Os | 77 Ir | 78 Pt | 79 Au | 80 Hg | 81 Tl | 82 Pb | 83 Bi | 84 Po | 85 At | 86 Rn |
| 87 Fr | 88 Ra | 89 Ac | 90 Th | 91 Pa | 92 U | 93 Np | 94 Pu | 95 Am | 96 Cm | 97 Bk | 98 Cf | 99 Es | 100 Fm | 101 Md | 102 No | 103 Lr | 104 Rf | 105 Db | 106 Sg | 107 Bh | 108 Hs | 109 Mt | | | | | | | | | |

Figure 1.2.1. The modern periodic table.

A more conventional, condensed representation of the periodic table, without atomic numbers, is shown in figure 1.2.2.

| H | | | | | | | | | | | | | | | | | He |
|---|---|---|---|---|---|---|---|---|---|---|---|---|---|---|---|---|---|
| Li | Be | | | | | | | | | | | B | C | N | O | F | Ne |
| Na | Mg | | | | | | | | | | | Al | Si | P | S | Cl | Ar |
| K | Ca | Sc | Ti | V | Cr | Mn | Fe | Co | Ni | Cu | Zn | Ga | Ge | As | Se | Br | Kr |
| Rb | Sr | Y | Zr | Nb | Mo | Tc | Ru | Rh | Pd | Ag | Cd | In | Sn | Sb | Te | I | Xe |
| Cs | Ba | La* | Hf | Ta | W | Re | Os | Ir | Pt | Au | Hg | Tl | Pb | Bi | Po | At | Rn |
| Fr | Ra | Ac† | Rf | Db | Sg | Bh | Hs | Mt | • | • | • | | | | | | |

| *Ce | Pr | Nd | Pm | Sm | Eu | Gd | Tb | Dy | Ho | Er | Tm | Yb | Lu |
|---|---|---|---|---|---|---|---|---|---|---|---|---|---|
| †Th | Pa | U | Np | Pu | Am | Cm | Bk | Cf | Es | Fm | Md | No | Lr |

Figure 1.2.2. The conventional periodic table.

The representation of the periodic table in figure 1.2.2 has all the same elements as the table in figure 1.2.1, but it requires the user to remember that the elements from cerium (Ce) to lutetium (Lu) and thorium (Th) to lawrencium (Lr) actually follow lanthanum (La) and actinium (Ac), respectively. In the conventional periodic table, these elements are shown as detached rows for typesetting convenience.

But whichever form one chooses to consult, the striking feature of the periodic table is its orderliness. In a world of shape-shifting clouds and randomly branching trees and no-two-look-alike cells or snowflakes, the order of the periodic table, rigid as an orthodox liturgy, may seem a bit unnatural. However, this primary order, dissolving through complexity into disorder (a discussion reserved for our chapter on thermodynamics), forms the fundamental basis for our varied world.

The first and most definitive pattern is in the number of protons in the nucleus of each element: hydrogen, the first element listed, has one proton in its nucleus. Helium, the second element (reading from left to right), has two protons in the nucleus. After helium, dropping down to the second row and starting again from left to right, lithium has three protons in the nucleus, and the next, beryllium, has four. Proceeding straight across, jumping the gap, boron is next and has five protons in its nucleus. The pattern continues: carbon has six, nitrogen has seven, oxygen has eight, and on it goes. The number of protons in the nucleus is called the *atomic number* and is the number given along with the element symbol in figure 1.2.1. Complexity starts to creep in, however, as soon as the mass of the elements is considered. The atomic numbers count up the elements in nice, neat whole numbers. But a look at the appendix tells us that the atomic *masses* for the elements are not nice, neat, whole numbers. The irregular nature of the atomic masses is a result of the fact that atoms are built from protons and neutrons. There is a basic theme that describes how they are built, but there are also variations on the theme.

You have to start somewhere, and when it came to mass, scientists started with the proton. They assigned the proton a mass of one *atomic mass unit*, or *amu*. Because they are so very close in mass, for many purposes the neutron is assumed to have a mass of one amu, too. The number of protons in an atom of an element is always the same and equal to its atomic number. If an atom has seventeen protons, then its atomic number is 17, and it is chlorine. If an atom is chlorine, it has seventeen protons. But the mass of a given atom of an element can vary because the mass of an atom is determined by the number of protons *and* the number of neutrons, and the number of neutrons can vary. For instance, each chlorine atom has seventeen protons, but some chlorine atoms have eighteen neutrons, and some have twenty neutrons. The total mass for the atoms of chlorine with eighteen neutrons will be seventeen amu plus eighteen amu, or thirty-five amu. The total mass for the atoms with twenty neutrons will be seventeen amu plus twenty amu, or thirty-seven amu. When there are atoms of an element that have different masses, these atoms are called *isotopes*. The atomic mass for a given element is the average mass of its isotopes.

The number of protons in an atom is sort of like the characteristic that distinguishes a human as male. If a person has this characteristic, then he is male. If an atom has a certain number of protons, then that decides what element it is. But different males can have different masses. We could easily find the average mass of a sample of males by adding all their individual masses and dividing by the number in the sample. The same procedure is followed for elements on the periodic table. As can be seen in the appendix, the mass for chlorine is given as 35.45 amu, which is not the mass of either isotope. It is an average over the natural abundance of isotopes. However, it is the number of protons that determines the element, so even though isotopes may have different numbers of neutrons and different masses, it is the atomic number that tells the number of protons, and the number of protons determines the element. The number that increases as you go across the periodic table is the atomic number, and this is the same as the number of protons.

But if the number of protons in the nucleus were the only basis of classification, then why would the table have its curious shape? The answer is that the order of the table goes deeper than the number of protons.

In biology, organisms are classified by kingdom, phylum, class, order, family, genus, and species. In chemistry, elements are also classified, though with fewer layers. What concerns us here is the classification of elements into what are termed different *species*. The number of protons determines the element, so atomic number could be thought of as the "genus." If an atom has two protons, then it is helium. If an atom is helium, then it has two protons. That doesn't change. The number of electrons in an atom of an element, however, can vary, and each variation can be thought of as a "species." In the neutral species of an atom, the number of protons is the same as the number of electrons.

Protons carry one unit of positive charge, so for an atom to be neutral, that positive charge must be offset by a negatively charged electron. Therefore, in addition to giving the number of protons in a particular element, the periodic table also tells us the number of electrons in an uncharged atom of the element. For instance, hydrogen has one proton, so a neutral atom also of hydrogen has one electron. Carbon has six protons in its nucleus, so it has six electrons in an electrically neutral atom.

The electrons occupy orbitals, as we previously outlined, and now we are going to add that these orbitals are arranged around the nucleus in *shells*. The shells are nested, as sketched in figure 1.2.3, and each one is able to accommodate more electrons. The number of electrons each shell can hold is determined by a mathematical progression: the first shell holds two electrons, the second shell holds eight electrons, the third shell holds eighteen electrons, and the fourth shell holds thirty-two electrons. A shell's capacity for holding electrons increases in a regular pattern: the square of the shell number times two. The first shell, shell 1, holds one squared times two, or two electrons. The second shell, shell 2, holds two squared times two, or eight electrons, and so on.

Here we have represented the shells as neat layers in nested spheres, sort of like a jawbreaker with many layers of flavor. But in truth, these

Figure 1.2.3. A pictorial representation of shells around an atom. Not to scale!

shells are made of subshells, and if we were to represent the subshells in our model, we would end up with something that looked more like a sliced-open onion. In an onion, the layers overlap and become interwoven, and that is true for the subshells of atoms, too. But this is a layer of the onion that we don't presently need. Theoretically, once filled with electrons, the shells are spherical, so we will consider them spherical here.[2]

The rows of the periodic table are called the *periods* (hence periodic table). The position that an element occupies in a period is determined by the number of electrons in its outermost shell. Each time the atomic number increases by one, the number of protons in the nucleus increases by one, and the new element requires one more electron for a neutral atom. The electrons arrange themselves in shells around the nucleus. Each time a shell is filled, a row is finished. A filled shell is represented on the periodic table as a filled row, as shown in figure 1.2.4.

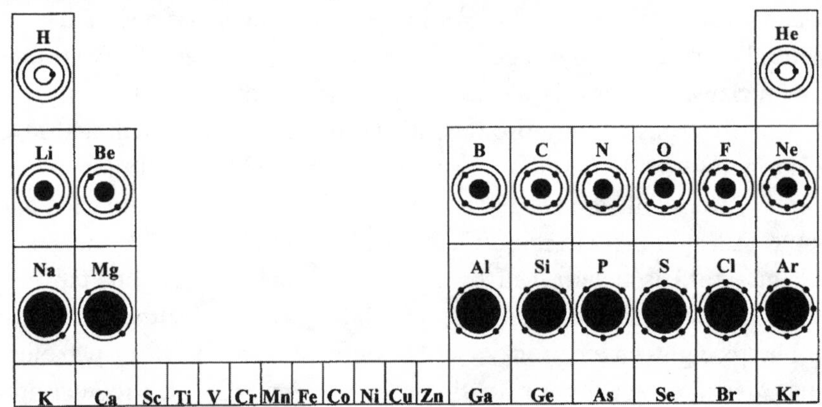


Figure 1.2.4. Each time the atomic number increases by one, the number of protons in the nucleus increases by one, and the new element requires one more electron for a neutral atom. The electrons arrange themselves in shells around the nucleus. Each time a shell is filled, a row is finished. Helium, neon, and argon each finish out a row, and each has a filled shell.

The shells, therefore, determine the shape of the periodic table, but there is more to the story than that. The existence of shells has another important effect: the occupancy or partial occupancy of a shell dictates the chemical behavior, or the *chemical reactivity*, of the various elements. The vertical columns of the periodic table are groups of elements that share a similar condition of occupancy of their outermost shell and are therefore members of the same family, but families that share a common chemistry rather than a common ancestry.

For instance, helium, neon, and argon each have filled shells, and this vertical column is commonly called the noble gases because of their dignified, inert, unreactive nature. Their lack of chemical reactivity is typified by helium, which is so unreactive that we use helium balloons for decorations at parties. Hydrogen, on the other hand, is very reactive. Hydrogen balloons at a party might result in entertainment more explosive than we would care for.

The chemical reactions of nitrogen and phosphorus are similar because they share the same number of electrons in their outer shell (five). The reactivity of oxygen resembles the reactivity of sulfur because of their shared outer-shell occupancy (six). This outer-shell occupancy of an atom is called its *valence*. Carbon has a valence of four (with four electrons in its outer shell), and its chemistry shares some similarities with silicon, which also has a valence of four. Silicon, germanium, tin, and lead, which have the same valence, have all been used in various proportions to form semiconductors, interesting and important materials that we will investigate later when we discuss chemical bonding.

Oxygen reacts with hydrogen in a ratio of one to two, as does sulfur, selenium, and tellurium. It is this chemical reactivity, the ability of elements to join chemically into compounds, that is responsible for transformation of the elements of the cosmos into the materials of Earth and ultimately the materials of life. Although we have to put off our detailed discussion of the complex molecules of life until we have a firmer foundation, we already have enough information to discuss two of the most important chemicals of life and decidedly life's necessary precursors: salt and water. The first life formed in the salty oceans, as evidenced by the salty solutions of our cells, and is always present in the salty character of our bodies.

Normally in the English-speaking world when we think of salt, we think of table salt, sodium chloride, NaCl. If there is a second-most familiar chemical formula, second to $H_2O$, that is, it must be sodium chloride, the formula for table salt. Sodium chloride and potassium chloride share so many properties that potassium chloride is used as a substitute for table salt. They have similar formulas, too, in that sodium chloride is formed from one part sodium and one part chlorine, and potassium chloride is formed from one part potassium and one part chlorine. But this ratio of one to one is not true for all salts. For instance, calcium chloride has a ratio of two parts chlorine to one part calcium. Why does sodium behave one way and calcium another? The periodic table speaks to that issue, too.

To begin with, the periodic table can be broken down into two major sections, the metals and the nonmetals. The metals are to the left of the zigzag staircase line that starts on the right between B (boron) and Al (aluminum) and ends between Po (polonium) and At (astatine). (With the exception of hydrogen, which is considered a nonmetal.) The metals Cu (copper), Ni (nickel), and Zn (zinc) are familiar and do a good job representing what we expect from metals: they are hard at room temperature, they can be bent and molded, and they conduct heat and electricity. The nonmetals are a bit harder to pin down with a set of generic characteristics, but the properties of oxygen and carbon are familiar enough to establish that these elements are *not* metallic in their behavior.

The division between metal and nonmetals is rather fuzzy, so the elements in the immediate vicinity of the zigzag staircase line are called *metalloids*, which means they don't fit either definition exactly. Aluminum (Al) finds itself in this position, and though we generally tend to think of aluminum as a metal, it is an odd metal because it is very light (that is why it is chosen for engine blocks in race cars) and because it can appear grainy and gray rather than sleek and shiny. Aluminum foil is malleable and shiny but changes drastically in the charcoal fire of a barbecue, as do aluminum cans. So aluminum sometimes acts like a metal and sometimes like a nonmetal. Elements like this can be called metalloids.

A salt is a compound formed from a metal and a nonmetal; thus, sodium, a metal, reacts with chlorine, a nonmetal, to form sodium chloride, the salt we know as table salt. The unvarying recipe for sodium chloride is one sodium atom with one chlorine atom, and the basis for this strict ratio is again shells, as we will now explain.

The mass of an atom is determined by the number of protons and neutrons in its nucleus, as we mentioned earlier. The electrons, you will recall, add virtually nothing to the mass, as fleas on an elephant add virtually nothing to the mass of the elephant. But just as fleas, despite their size, influence the behavior of the elephant, the electrons, despite their small mass, determine the chemistry of the elements. When an electron is gained or lost, the atom experiences an imbalance of charge and becomes an *ion*, which can be considered a new species of that element. An atom of an element with more electrons than protons is a negatively charged species, a negative ion. An atom of an element with fewer electrons than protons is a positively charged species, a positive ion. An ion is a reactive species, seeking, as it were, an entity of the opposite charge with which to combine.

It might seem at first that any amount of charge could be gained or lost and any ion could be formed, but it turns out that there are certain predictable amounts of charge that each element prefers to handle—sort of a comfort level for ionization, if you will. This comfort level can be predicted from the periodic table.

The first shell (the first row) is filled with two electrons; the second shell needs eight; the third shell holds eighteen; and so on. We said that the noble gases were unreactive, and it turns out that they also have filled shells. The first and most fundamental principle of the *shell theory of atoms*—the theory that chemical behavior depends on the number of electrons in shells—is that atoms like to have fully filled shells. A demanding motivation for chemical reactions is for atoms to fill or empty their electron load, by whatever means available, until they reach this happy state. Thus, chlorine (Cl), with seven outer electrons, will tend to acquire an electron so that it can have eight electrons and fill its shell. Sodium (Na) will shed its outer electron so that it can have just a filled shell of eight. In fact, the number *eight* is so often the choice for stable electron arrangements, we refer to all these stable, filled-outer-shell arrangements as *octets.* Hence musicians have their octaves, and chemists have their octets. But for chemists the concept is a bit fuzzier. For hydrogen, for instance, a filled outer shell consists of two electrons, so its "octet" is two. Actually a duet.

Shedding an electron gives sodium a net charge of positive one, which is shown as a superscript + on the symbol for the element, $Na^+$. Of course, sodium could also add seven electrons to fill its octet, but atoms like to do what is easiest. Losing one electron is easier than taking on seven more. Adding an electron gives chlorine a net charge of negative one, which is shown as a superscript – on the symbol for the element, $Cl^-$. The equal and opposite charges of the sodium ion and the chloride ion mean that the two will seek each other out and combine in a one-to-one ratio, NaCl.

| H | | | | | | | | | | | | | | | | | He |
|---|---|---|---|---|---|---|---|---|---|---|---|---|---|---|---|---|---|
| Li | Be | | | | | | | | | | | B | C | N | O | F | Ne |
| Na | Mg | | | | | | | | | | | Al | Si | P | S | Cl | Ar |
| K | Ca | Sc | Ti | V | Cr | Mn | Fe | Co | Ni | Cu | Zn | Ga | Ge | As | Se | Br | Kr |
| Rb | Sr | Y | Zr | Nb | Mo | Tc | Ru | Rh | Pd | Ag | Cd | In | Sn | Sb | Te | I | Xe |
| Cs | Ba | La* | Hf | Ta | W | Re | Os | Ir | Pt | Au | Hg | Tl | Pb | Bi | Po | At | Rn |
| Fr | Ra | Ac† | Rf | Db | Sg | Bh | Hs | Mt | • | • | • | | | | | | |

| *Ce | Pr | Nd | Pm | Sm | Eu | Gd | Tb | Dy | Ho | Er | Tm | Yb | Lu |
|---|---|---|---|---|---|---|---|---|---|---|---|---|---|
| †Th | Pa | U | Np | Pu | Am | Cm | Bk | Cf | Es | Fm | Md | No | Lr |

Figure 1.2.5. If sodium, Na, loses an electron, it will have a filled shell. If chlorine gains an electron, it will have a filled shell. With one fewer electron, sodium becomes $Na^+$. With one extra electron, chlorine becomes $Cl^-$. $Na^+$ and $Cl^-$ can come together to form NaCl.

It works the same way for potassium chloride. Potassium (with the symbol K for the Latin *kalium*) is right below sodium on the periodic table. Potassium can attain a blissful state of eight electrons in its outer shell by losing one electron, which leaves one extra, unchecked proton and a positive charge. This one positive charge matches with the negative charge of the chloride ion in its happiest state, and the result is KCl, a compound able to substitute for table salt. Consulting the periodic table, we see that calcium realizes a filled-shell configuration when it loses two electrons and acquires a plus-two charge, $Ca^{2+}$, so, predictably, it will join with two chloride ions, 2 $Cl^-$, to form the salt $CaCl_2$.

$$Ca^{2+} + 2\ Cl^- \rightarrow CaCl_2$$

So there you have it in a nutshell. Or an atomic shell, if you wish.

The idea can be extended to materials formed from nonmetals, too. For instance, the formula $H_2O$ can be explained in terms of atomic shells.

There are some subtleties regarding the difference between the bonding in sodium chloride and the bonding in water that we will explore later when we delve more into chemical bonding, but the rearrangement of electrons to fill shells is still our basic premise.

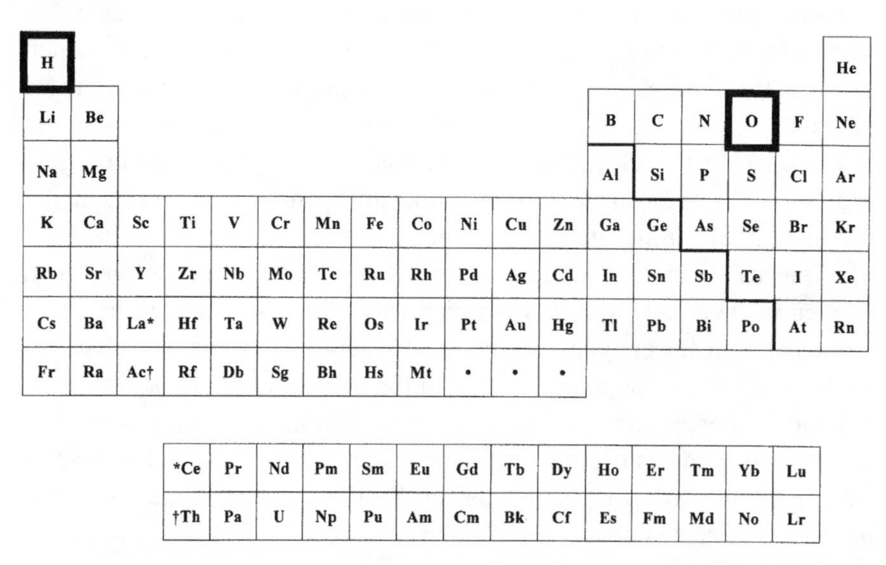


Figure 1.2.6. The number of valence (outer-shell) electrons for hydrogen and oxygen can be determined from their position in the periodic table. Hydrogen has an atomic number of 1, so it has one proton and one valence electron. Oxygen has a total of eight electrons, but two of these are in a filled shell. Only the outer six electrons are found in the valence shell.

The number of valence electrons (outer-shell electrons) brought into the mix by hydrogen is one, as evidenced by its position as the first element on the periodic table. Hydrogen craves a filled shell, such as that enjoyed by helium, which means it needs another electron. Similarly, the number of valence electrons brought in by oxygen is six, as evidenced by its position as the sixth element in period two, reading from left to right across the periodic table. However, oxygen is not satisfied, either. True happiness (stability) for oxygen would be to have a filled shell, as in the case of neon. To fill its shell, however, oxygen needs two more electrons. Hydrogen and oxygen fulfill their mutual needs by sharing electrons, but in this particular mélange, two hydrogen atoms are required to fill the needs of one oxygen atom. When the two hydrogen atoms share their two lone electrons with oxygen, oxygen ends up with eight valence electrons. In the exchange, each hydrogen is able to gain the use of one electron from oxygen and fill its own shell . . . and destiny. The resulting arrangement is water, $H_2O$.

There can be no question that we need water for life, and there is also no question that we need air. But not every component of air. Air is mostly nitrogen gas, which exists as the *diatomic* (two-atom) molecule $N_2$. Mixed in with nitrogen is a substantial amount of oxygen gas, which exists as the diatomic molecule $O_2$. The natural existence of nitrogen as a diatomic can also be explained by the desire to have filled shells. An atom of nitrogen has five electrons in its outermost shell. In diatomic nitrogen, each shares three electrons with its fellow nitrogen—which gives both the use of eight electrons—to fill their shells.

The reason the atmosphere is much more enriched in nitrogen than oxygen is chemical. Nitrogen is much less chemically active; that is, it participates in fewer types of chemical reactions than oxygen does. In addition, nitrogen gas is very stable in the atmosphere, exposed to solar radiation, whereas oxygen is not. But to say that nitrogen gas is less reactive than oxygen gas is not to say the element of nitrogen is unreactive. It is, in fact, very reactive, as evidenced by the number of materials and mixtures that have nitrogen as a major constituent.

For instance, nitrogen is a principle component of gunpowder, yet nitrogen is also an essential component of amino acids and necessary for life as we know it. Nitrogen compounds give smog its red-brown color, nitric acid is found in acid rain, and liquid nitrogen is used to remove warts. The same nitrites that preserve food cause cancer in mice. Nitrogen compounds are responsible for nitrogen-rich, fertile soil and also for the growth of diaper-rash bacteria. Some compounds of nitrogen are powerful cleaning agents, and others are the source of the fishy odor of degrading fish.

The laughing gas that enables painless tooth extraction is an oxide of nitrogen, but TNT stands for trinitrotoluene, a less constructive compound. Physiologically active nitrogen compounds such as caffeine, nicotine, morphine, codeine, and quinine have shaped human history. The chemistry that threatens life with nitrogen-based fertilizer bombs is shared with the chemistry that protects life through the cushion of nitrogen air bags. So destructive and so nurturing at the same time. How can this be?

To understand the virtuosity of nitrogen, we need to take another look at the periodic table. From nitrogen's position we can see that there are two ways it can achieve a filled shell: it can gain three electrons to move forward to a filled shell, as noted above, or it can also lose five electrons to go backwards to a filled shell. As it turns out, there are some intermediate situations, such as losing only three electrons, acceptable to nitrogen, too; therefore, nitrogen has several personalities and is able to undergo many metamorphoses. We will continue to learn about the versatility of nitrogen—and the rest of the periodic table—in the pages to come.

## For Example: Elements of Diversity

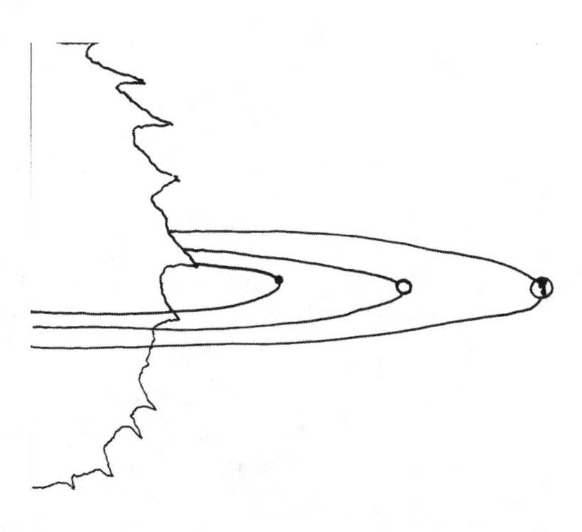

Now that we've been introduced to the periodic table, let's take a short guided tour. Many of the elements on the periodic table, such as iron (Fe, from the Latin name for iron, *ferrum*), nitrogen (N), phosphorus (P), oxygen (O), and carbon (C), are familiar. Others, such as technetium (Tc) and lanthanum (La), generally are not. We are familiar with hydrogen and oxygen because these elements are evident in the air we breathe and the dirt we walk on, but no matter how we pant or how we dig, we do not expect to ever inhale one atom of lawrencium (Lr) or unearth one gram of seaborgium (Sg). The unfamiliarity of these elements is not just based on difficulty in pronunciation and spelling (though that could be argued), but on cosmology and physics. The reason has to do with something called the Big Bang.

According to the Big Bang theory, our universe was born with an unimaginably huge explosion (hence the name Big Bang). The debris of the Big Bang first condensed into the lighter elements, hydrogen and helium, because they are relatively simple: hydrogen is composed of one proton and one electron, and helium is composed of two protons, two neutrons, and two electrons. Only when enough of these atoms had gathered together and the mass started to compress under its own gravity did these nuclei begin to fuse into the nuclei of heavier atoms. The heaviest nucleus that could have been formed in the first stars by the *fusion* process, the fusing together of nuclei, is iron. To contain the elements it has, our sun must have begun its fusion process with some heavier nuclei already present, which means it is probably a third-generation star—the result of the collected dust from the explosions of two previous stars. But even so, a look at the relative abundance of the elements in the sun, as shown in figure 1.2.7, reveals that the lightest elements, hydrogen and helium, are still the clear winners.

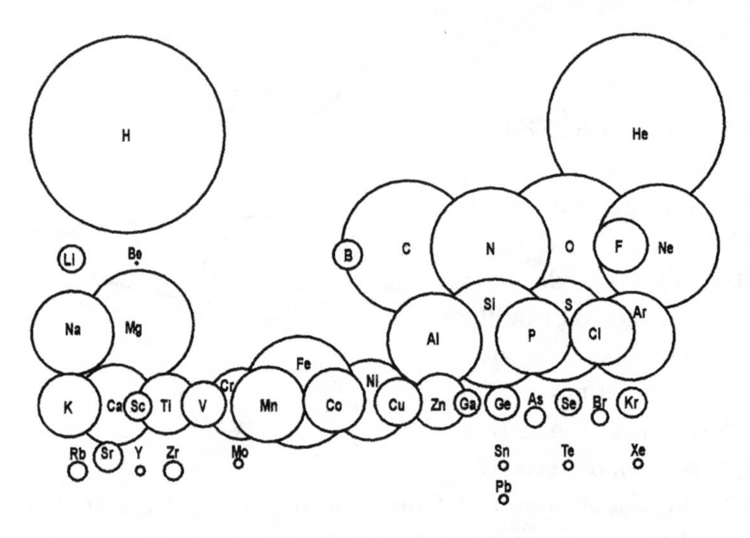


Figure 1.2.7. The abundance of elements in the sun. Adapted with permission from Steven I. Dutch, "Periodic Table of Elemental Abundance," *Journal of Chemical Education* 76 (1999): 356–58.

In figure 1.2.7, the abundances are displayed with a *logarithmic scale*: the changes in the radius of the circle for the element represent changes of orders of magnitude, or powers of ten, like a decibel scale. A sound registering 20 on the decibel scale is ten times as loud as a sound registering 10; a sound registering 30 is one hundred times greater than a sound registering 10. On this periodic table, it is the radius of the circle around the element that represents the multiple of 10.

We use a logarithmic scale for the abundance of the elements because there is so much difference between them. An attempt to compare abundances without this device would be a bit like trying to picture Godzilla and a mouse in the same movie frame (which may be the reason there is no *Godzilla versus Mighty Mouse* movie). A skyscraper can be shown along with Godzilla, but not a mouse. We will not concern ourselves with exact numbers here; we just need a feel for the relative abundances of the elements, and these are nicely represented by the circles.

After hydrogen and helium, there is a jump to carbon because intermediate nuclei are too unstable to form in the fusion process. Helium has two protons in its nucleus and carbon has six, so a carbon nucleus can be made from the fusion of three helium nuclei, a process that requires the extreme heat and density of massive stars. After that there is a marked preference for nuclei with an even number of protons in the nucleus, oxygen has eight protons, neon has ten protons, magnesium has twelve protons, and so on. This preference can be explained by the theories of nuclear physics—but that is a topic for another time. We note that the elements essential to life, hydrogen, carbon, nitrogen, oxygen, sodium, magnesium, phosphorus, calcium, iron, and so forth, are all there in the sun, too.

Sometimes the flotsam of the Big Bang and exploding stars can condense into a planet rather than another star, which is probably what happened in the case of our planet, Earth. From the abundance of elements on our Earth's crust shown in figure 1.2.8, it can be seen that the material of Earth is essentially that of the sun, depleted in the more volatile elements such as helium and hydrogen.

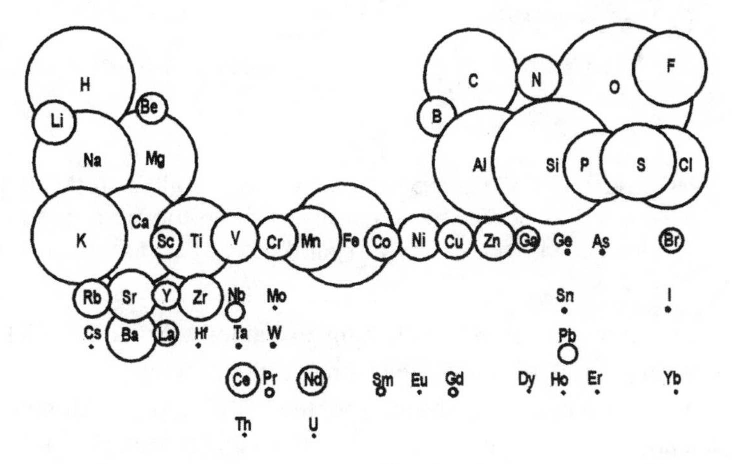


Figure 1.2.8. The abundance of elements in Earth's crust. Adapted with permission from Steven I. Dutch, "Periodic Table of Elemental Abundance," *Journal of Chemical Education* 76 (1999): 356–58.

The relative wealth of aluminum and silicon compared to iron and nickel shows a shift in favor of the lighter elements because we are looking at Earth's crust: in the upheavals that have occurred over the eons, the heavier elements have sifted toward the interior and the lighter elements have risen to the top.

In figure 1.2.9, we have compared the top forty elements in Earth's crust with the top ten elements in the human body.[3] Reassuringly for most people, though perhaps discouraging for theorists of extraterrestrial seeding, it can be seen that the abundance of elements in the human body is remarkably similar to the abundance of elements on Earth, which rather nails down our evolutionary origin. In order of decreasing amounts, the human body is composed of oxygen, carbon, hydrogen, nitrogen, calcium, phosphorus, potassium, sulfur, sodium, and chlorine. Other elements are present, but in quantities of less than 0.1 percent.

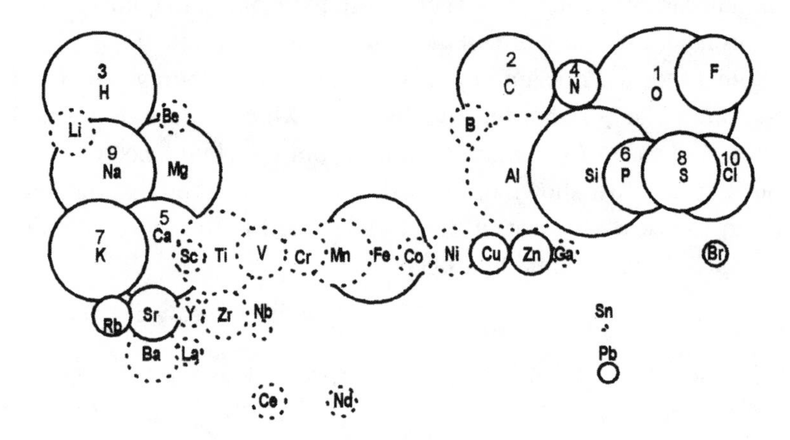


Figure 1.2.9. Comparison of the top forty elements in Earth's crust with the top ten elements in the body. The elements found in the body are numbered to show their abundance. Oxygen is the most abundant (1), carbon is second (2), etc.

There are, however, some interesting exceptions. Silicon, which is present in every grain of sand, tread on daily by every creature, and gripped firmly by every root, is absent from the body's top ten. Aluminum, also conspicuous in Earth's crust, not only is missing from the body's top ten but also is being investigated for its links with neurological damage.[4] Iodine doesn't make the top forty for abundance in Earth's crust, but if it is absent from the body, people suffer goiters. From this comparison we see that it is not just physical presence that decides the importance of an element but its chemical reactivity, too—so into that realm we proceed.

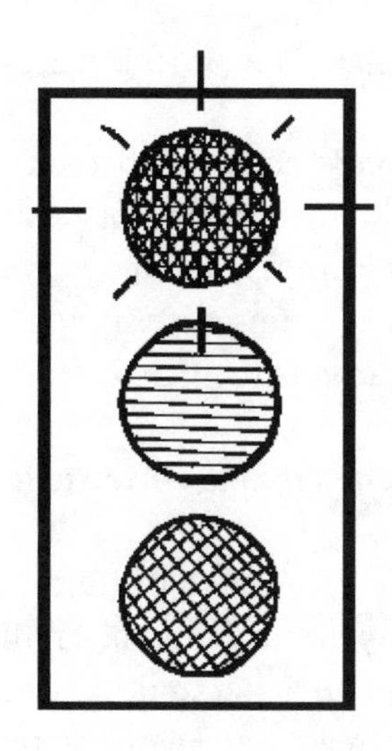

## Demonstration 3: Stop-and-Go Chemistry

*Our earthly fire also consumes more or less rapidly according as the object which it attacks is more or less combustible so that human ingenuity has even succeeded in inventing chemical preparations to check or frustrate its action.*

—James Joyce, *Portrait of the Artist as a Young Man*, ca. 1915

Don the safety glasses. Ladle two tablespoons (30 milliliters) of iron acetate solution (the vinegar solution with dissolved steel wool described in the "Shopping List and Solutions") onto a plastic-coated paper plate. Add a teaspoon (5 milliliters) of household ammonia. The resulting solution should turn from a light orange-brown to a dark green. Now drop in a teaspoon (5 milliliters) of hydrogen peroxide. The resulting solution is a deep blood-clot red.

What happened? Iron, like nitrogen, can tolerate different numbers of electrons associated with its nucleus. The *oxidation state* of an element in a given environment—be it a free element, an ion, or in a compound—is an expression of the element's electron load in that environment. In the current demonstration, iron is changing between two different oxidation states: the *ferric ion* ($Fe^{3+}$, or iron with a positive three charge, which means it is missing three electrons) and the *ferrous ion* ($Fe^{2+}$, or iron with a positive two charge, which means it is missing two electrons). The agent causing the change in the iron is the peroxide.

ferrous iron + hydrogen peroxide → ferric iron + water + oxygen

Iron is changed from the ferrous ion to the ferric ion because the peroxide attracts the electrons away from the iron. Hydrogen peroxide uses the extra electrons to change into water and oxygen gas, which bubbles out of the solution during the reaction. The compound formed between ferrous iron and ammonia is green. When the peroxide changes the ferrous iron to the ferric iron, the ammonia, still in solution and unchanged, forms a red compound with ferric iron. Reactions such as these, where elements gain or lose electrons, are called *reduction* (if they gain electrons) or *oxidation* (if they lose electrons). Because the reactions must happen in tandem, the combined result is called a *redox reaction*, a class of reactions of current and historical importance.

# CHAPTER 3

## Reason, Reactions, and Redox

*They first broke the ore into little pieces, and cleansed them with the hand. . . . Then coal and ore were arranged in heaps and in successive layers. . . . In this way . . . the coal would be transformed into carbonic acid, then into oxide of carbon, its use being to reduce the oxide of iron, that is to say, to rid it of the oxygen.*

—Jules Verne, *Mysterious Island*, ca. 1870

*Doubtless a vigorous error vigorously pursued has kept the embryos of truth a-breathing: the quest of gold being at the same time a questioning of substances, the body of chemistry is prepared for its soul, and Lavoisier is born.*

—George Eliot, *Middlemarch*, ca. 1871

Like a fickle lover, an electron will flirt with, or even flit to, any nucleus it finds more attractive. In the last demonstration, "Stop-and-Go Chemistry," we saw electrons leave iron quite cavalierly in favor of hydrogen peroxide. True, the electrons were content to remain with iron as long as there were no better prospects in the vicinity, but in the presence of hydrogen peroxide, away they went. In other words, the degree of association, lover with beloved, electron with nucleus, depends on the environment. The ferrous ion is iron missing two electrons; the ferric ion is iron missing three electrons. The green ferrous ammonia compound became a red ferric ammonia compound because hydrogen peroxide acts like an electron sponge.

When an element loses electrons, it is said to be *oxidized*. In the case of our stoplight reaction, the moniker makes sense: oxygen is acting on the iron, taking its electrons. However, for other reactions, *oxidation* is a misnomer: oxygen does not have to be involved for oxidation to take place. For instance, with safety glasses in place, take a metal paper clip (not plastic coated), unwind it part way, and put the unwound end in your copper sulfate solution. In five minutes or so, the end of the paper clip in solution will have acquired a shiny copper coating. In this particular case, copper ions in solution are gaining electrons lost from the metal of the paper clip and turning into copper metal, an exchange of electrons that occurs without the intervention of oxygen.

When an element gains an electron, it is said to be *reduced*. The first quotation at the beginning of this chapter, the quote from Jules Verne, describes just such a process: the gleaning of iron from iron ore. Metal smelting was one of the first organized and controlled chemical reactions (after, perhaps, cooking, pottery, painting, and fermentation). Metal smelting is the process for which the gaining of electrons, or *reduction*, is named. The process is called reduction because the metal smelted from the ore weighs less than the beginning ore; that is, it is reduced in weight.

Most metals exist in nature combined with one or more nonmetals, such as oxygen, sulfur, or chlorine. They are compounds in ores, not pure metals. In the smelting process, the metal is separated from the nonmetal by heating the ore with a material that has a stronger attraction for the nonmetal than the metal. In the case of the process described by Jules Verne, the nonmetal is oxygen and the second material with a stronger attraction for oxygen is carbon, provided in the form of charcoal. The oxides of carbon, such as carbon dioxide, are mostly gases that will blow away, leaving the pure metal behind. Along with the understanding of atomic structure came the realization that the metal had gained electrons in the process of being reduced from ore to metal, so *reduction* became the name for gaining electrons. Charcoal's experience is called *oxidation* because the carbon combines with the nonmetal to carry it off, and many times this is oxygen. At the atomic level the carbon loses electrons, so losing electrons came to be known as oxidation. The defining characteristic of a redox reaction is that the element being reduced gains electrons and the element being oxidized loses electrons, a convention that has been immortalized in the mnemonic "LEO the lion says GER," which stands for "Lose Electrons Oxidation" and "Gain Electrons Reduction." Whenever there is oxidation there must be reduction; that is, the reaction pair is always found together: you can't have something losing electrons without the electrons going someplace else.

The copper sulfate and iron acetate solutions made from the recipes in the "Shopping List and Solutions" are examples of redox reactions. In the copper sulfate solution, the copper metal loses two electrons to become copper ion, $Cu^{2+}$, so it is oxidized. The fumes are from the nitrogen/ oxygen compounds in salt petre being reduced to volatile nitrogen/oxygen compounds. In the iron acetate solution, the iron in the steel wool is oxidized to $Fe^{2+}$ (recall that the symbol for iron is Fe from *ferrum*). The electrons given up by iron are taken up by the hydrogen ion in vinegar to form the hydrogen gas bubbles that you probably observed in the iron acetate solution.

The fact that most metals exist in nature in a combined form (combined with nonmetals) can be satisfactorily demonstrated by a tour of a beach or a rocky field. One finds sand and rocks in all shades and textures, but unless one is very lucky, no chunks of pure metal. Not that it is impossible to find pure metal. Even nuggets of gold, as evidenced by the California gold rush, are possible to find. But such occurrences are rare, which helps give gold its value. The desire to transmute more common metals into precious gold was one of the goals of alchemy—an early, unsystematic practice of chemistry that flourished in Alexandrian Egypt and later in Europe during the Middle Ages. Based on colorful transformations such as the ones witnessed in our "Stop-and-Go" demonstration, alchemists believed that if they could find the right recipe, they would be able to transform other metals into gold. George Eliot alluded to this quest in the quotation from *Middlemarch* at the beginning of this chapter. As she implies, no alchemist ever found a way to make gold. But the effort led to many fundamental revelations, such as the discovery of oxygen as a pure element. This discovery was made by several scientists at about the same time, including Antoine Lavoisier, the French chemist referred to by George Eliot as the "soul" of chemistry. But while Lavoisier may have been born to be the soul of chemistry, he died on the guillotine during the French Revolution. There may be joy in chemistry, but the same cannot always be said for politics.

As our quick trot from prehistoric chemistry to the French Revolution might indicate, paired reduction-oxidation reactions, or redox reactions, have played a continuing role in the history of humankind. In our one-sentence survey of prehistoric chemistry, we did not include an extremely important redox reaction—the making of fire—but we include it now because combustion, it turns out, is one of the most famous, and infamous, redox reactions.

## For Example: Firefighters and the Chemistry of Combustion

Combustion is a fascinating reaction: from the gentle light of a candle, to the comfort of the hearth, to the welcoming warmth of a campfire. But these are cases where combustion is controlled. In the cases of uncontrolled combustion, the fascination turns to fear, primal and pure. Thankfully, there are those who routinely face this fear to protect others from the danger. A brief consideration of combustion chemistry hints at the magnitude of their challenge.

The primary combustion reaction is a straightforward redox reaction. A compound made solely of hydrogen and carbon, a *hydrocarbon*, is usually the fuel. When hydrocarbons mix with oxygen, they can react to produce carbon dioxide ($CO_2$) and water and heat.

$$\text{Hydrocarbon} + \text{oxygen} \rightarrow \text{carbon dioxide} + \text{water} + \text{heat}$$

In the reaction, the carbon is oxidized (literally gaining oxygen this time) and the oxygen gas gains enough electrons to attract hydrogen and form water, so it is reduced.

Gasoline is a mixture of hydrocarbons. The reactions in the cylinders of cars are combustion reactions in which hydrocarbons and oxygen are combined, as in the above equation, so car exhaust, the reaction product, consists primarily of carbon dioxide and water. Sugar and starch are relatives of hydrocarbons (they have C, H, and O), oxygen is in the gas we breathe, so heat-producing metabolism can be thought of as a combustion reaction. That metabolism is a form of combustion can be demonstrated by blowing out a candle: the rush of oxygen-poor and carbon dioxide–rich air both cools the flame and deprives it of oxygen. Wood, paper, and dry grass are mainly hydrocarbon derivatives; they make excellent fuels and dangerous fires. The consumption of oxygen by fire and the production of carbon dioxide make suffocation as much of a hazard to the firefighter as thermal injury.

That oxygen is necessary for combustion can be demonstrated by covering a candle with a nonplastic clear glass. After all the oxygen in the glass is used up, the candle goes out. Showing that carbon dioxide is a product of combustion is a bit more complicated and requires safety glasses and gloves, but it can be done. Take two similar nonplastic (real-glass) glasses and to one add four teaspoons (20 milliliters) of water. Add two drops of the fish-tank indicator suggested for purchase in the "Shopping List and Solutions." Placing both glasses on a sheet of white paper for viewing, you should see that the one with water has a faint green-blue color.

Light the candle and get a good flame going. Invert the empty glass over the candle. Allow the flame to burn out. Take the glass off the candle and immediately pour the water-indicator solution from the other glass into the glass that was over the candle. Quickly cover the mouth of the receiving glass with your palm and swirl the solution a bit so it can mix with the gases from the candle flame. Now put this glass back on the white paper and observe the color. The $CO_2$ in the flame's gases should have changed the indicator from green-blue to yellow.

How do we know that it was $CO_2$ that caused the change? Take a clean glass and prepare another solution with four teaspoons water (20 milliliters) and two drops of fish-tank indictor. Now blow up a balloon. Holding the balloon pinched shut, put the balloon on one end of a straw and put the other end of the straw down into the indicator solution. When you ease open the neck of the balloon, freshly exhaled $CO_2$ from your lungs will bubble quickly through the solution and you should see the indicator once more turn from green-blue to yellow.

That water is a product of combustion can be demonstrated with an ice cube and a nonplastic glass or a nonplastic saucer. Invert the glass over the candle flame for a minute with the ice cube balanced on top of the glass, on the outside. Moisture from the flame will condense on the inside of the glass. Alternatively, place the ice cube on the saucer and then hold the saucer over the flame without touching the flame. Moisture will condense on the bottom.

Moisture can sometimes be observed on a metal spoon held over a flame, but water will not be the only material observed on the spoon. There will also be soot. Combustion is usually not the clean, simple reaction outlined above. Combustion is usually a complicated conglomerate of many reactions in which fragments of hydrocarbons, or soot, are produced. Some of the most interesting behaviors of combustion reactions are a result of incomplete combustion and more complicated reactions.

For instance, fire can spread, almost like a living thing, in pulsating waves. Part of the propagation is due to falling or flying debris, but because flames are burning gases, these gases and fire can travel anywhere the wind goes. To show that the flame contains burning gases, try the following experiment. Find an area that is free of stray breezes and clear the area of any flammable material (including loose hair and clothing). Put on your safety glasses. Light a candle with a wooden match and then hold the match aside, still burning. Blow out the candle and quickly bring the burning match close to the candlewick, but without touching the wick, as shown in figure 1.3.1. The flame will jump from the match to the candle as it races back through the combustion gases. You may have to try it a couple of times to get it right, but it should work.

Jumping through the air and spreading by falling debris aren't the only ways fires can propagate. Differences in air pressure are associated with fires because they consume oxygen and produce heat. The consumption of oxygen creates a partial vacuum at the source of the flame, which is filled by air coming in from the sides. The heated air above the fire causes an updraft. All forces together conspire to create considerable convection currents and pressure differences. Just as pressure differences in the atmosphere can produce tornadoes, pressure differences in a fire can cause what is called a *blowup*: a whirl of fire that can travel at nearly the speed of the wind.

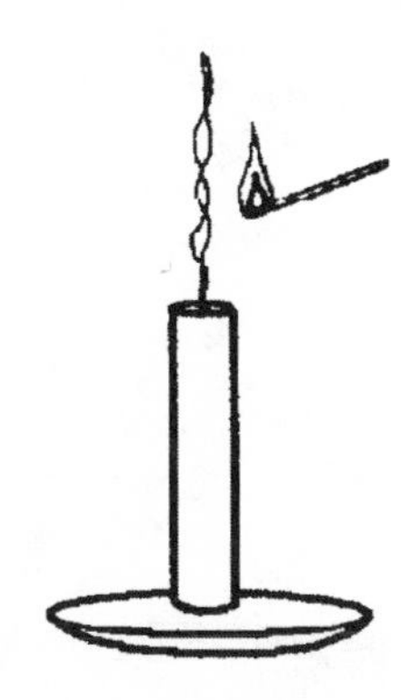

Figure 1.3.1. Bring a lit match up to a recently extinguished candle wick and the flame will "jump" through the burning gases and relight the wick.

In his excellent monograph *Young Men and Fire*, Norman Mclean states that these natural phenomena are poorly understood because there have been few reported direct observations: to compile a report the observer would have to be a survivor and a survivor would be someone who did not wait around to observe.[1] As Mclean details, however, one person, a parachuting firefighter, or *firejumper* as they are called, managed to survive such a whirling fire. He burned a bare patch ahead of the blowup and buried his face in the ashes as the blowup passed over, an event that took five minutes by his watch. He reported being picked up off the ground several times by the vortex but was able to maintain minimal breathing because of oxygen that remained close to the ground. That oxygen was necessary for his bodily combustion—the redox reaction that had to continue if he were to return to tell the story.

As pervasive as redox reactions are, however, they are certainly not all there is to chemistry. Another very common and important type of reaction is found in the acid-base reactions that occur in the air, on skin, in stomachs, in sewers, and even in swimming pools and fish tanks. The fish-tank indicator that we used to detect carbon dioxide is an acid-base indicator. It worked to detect carbon dioxide because carbon dioxide dissolved in water turns water slightly acidic. Intrigued? Read on.

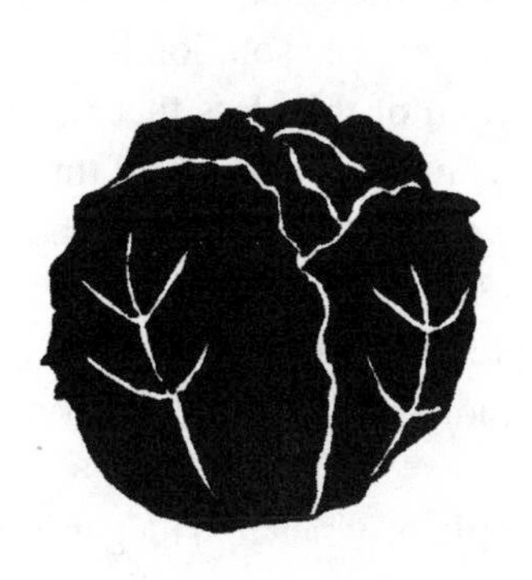

# Demonstration 4: Purple-Cabbage Indicator

*Tragedy is like strong acid—it dissolves away all but the very gold of truth.*

—D. H. Lawrence, *The Letters of D. H. Lawrence*, ca. 1920

It has been known since at least the 1600s, and probably much earlier, that some highly colored plants and flowers, such as red cabbage, contain a dye that will assume various colors in different chemical environments. Such a dye is called an *indicator* because its color indicates its environment. The purple-cabbage indicator is used to indicate whether a solution is acidic or basic. Briefly, acid solutions are sour solutions such as lemon juice and vinegar. Basic solutions are caustic and bitter like soapy water and household ammonia.

To see how this works, put on your safety glasses and then add about a tablespoon (15 milliliters) of purple-cabbage indicator (prepared as outlined in the "Shopping List and Solutions") to a cup (240 milliliters) of water and observe the color. It should be light lavender. Now make up a saturated solution of baking soda in water. A saturated solution is one in which no more solid can be made to dissolve. Baking soda is not very soluble in water, so it should not take more than two tablespoons (30 milliliters) of baking soda per cup of water (240 milliliters). The solution is saturated when there is baking soda left on the bottom of the glass that will not dissolve. Add purple-cabbage indicator to this mixture and it will turn blue. Baking soda dissociates into two ions in water, the sodium ion and the *hydrogen carbonate ion*, otherwise known as *bicarbonate*. The sodium ion is neutral, but the bicarbonate is a base. Purple-cabbage indicator turns blue in baking soda solution; therefore, blue indicates a basic solution when using purple-cabbage indicator.

Add some purple-cabbage indicator to a sample of vinegar. Vinegar is an acid. The vinegar turns the cabbage indicator bright pink, so pink indicates an acid solution when using purple-cabbage indicator.

Using one-cup (240-milliliter) samples of saturated baking soda solution, vinegar, ammonia, and water, you can set up the following banks of solutions and observe the colors that result when you add purple-cabbage, swimming-pool, or fish-tank indicator.

| Swimming-pool phenol red indicator | |
|---|---|
| water | reddish orange |
| vinegar | yellow |
| ammonia | reddish violet |
| baking soda | pink |

| Fish-tank bromothymol blue indicator | |
|---|---|
| water | greenish blue |
| vinegar | pale yellow |
| ammonia | pale blue |
| baking soda | blue |

| Purple-cabbage indicator | |
|---|---|
| water | purple |
| vinegar | shocking pink |
| ammonia | green |
| baking soda water | teal blue |

Now take some of the clear, carbonated soda that was a suggested purchase in the "Shopping List and Solutions" and add the purple-cabbage indicator. You should observe a pale pink solution. Carbonated soda is slightly acidic because carbon dioxide dissolves in water to form carbonic acid. Now take the baking soda/purple-cabbage solution and slowly and carefully pour the blue liquid that is over the undissolved baking soda (a process called *decanting* by chemist and wine connoisseur alike) into the carbonated soda/purple-cabbage solution. Eventually the solution should have a color very close to that for indicator in pure water. Why? Because acid *neutralizes* base and vice versa. What's neutral? Water. What was the old-time remedy for stomach acid? Baking soda.

Cheers!

# CHAPTER 4

## The Basic Stuff

> *"Ask Asia for a cup of sour cream, then your cakes will be light without much soda, which I don't like," was the first order. Demi tore downstairs, and returned with the cream, also a puckered-up face, for he had tasted it on his way, and found it so sour that he predicted the cakes would be uneatable. Mrs. Jo took this occasion to deliver a short lecture from the stepladder on the chemical properties of soda, to which Daisy did not listen, but Demi did, and understood it, as he proved by the brief but comprehensive reply: "Yes, I see, soda turns sour things sweet, and the fizzling up makes them light. Let's see you do it, Daisy."*
>
> —Louisa May Alcott, *Little Men*, 1871

In the above quote, Demi's sour-sweet reaction is an acid-base reaction. Baking soda is a base (as we found in the purple-cabbage demonstration), and sour cream is an acidic mixture: sour cream is made sour by the acid products of bacterial action. In the bottle rocket demonstration, we saw another acidic solution, vinegar, encounter baking soda and experience an explosive reaction. This same type of reaction, on a more modest scale, is what makes Daisy's cakes light:

$$\text{acid} + \text{baking soda} \rightarrow \text{carbonic acid} \rightarrow \text{carbon dioxide (a gas)} + \text{water}$$

Acid reacts with baking soda, and their product, carbonic acid, quickly turns into carbon dioxide and water. Carbon dioxide is the gas that blew the cork out of the bottle rocket and that blows bubbles into our fluffy little cakes. The reason you were cautioned to buy baking soda in the "Shopping List and Solutions," rather than baking powder, is because baking powder contains an acidic compound such as cream of tartar. Cream of tartar acts as a *leavening agent*; that is, when it dissolves in water with baking soda they come together and react to form bubbles.

Not all acid-base reactions create such volatile products. Vinegar is an acid, as indicated by the pink color of its solution with purple-cabbage indicator, and ammonia is a base, as indicated by the blue color of its solution with the purple-cabbage indicator. But when vinegar and ammonia are poured together without added indicator, there are none of the usual signs of a chemical reaction: no new fumes, no frothing, no heat, no light, no color change, no solid formation. If it weren't for the sour odor of vinegar and the pungent odor of ammonia, it would appear that two clear, watery liquids were being poured together. However, if a teaspoon (5 milliliters) of purple-cabbage indicator is added to a cup (240 milliliters) of each before pouring them together, the result will be a change in the color of the solution. Indicators are used to show evidence of a chemical reaction when there may be no other obvious sign.

So what are acids and bases? Vinegar is actually a dilute solution of acetic acid in water, about a 5 percent solution, but it rather nicely displays the characteristic properties of acids: they are sour, they turn purple-cabbage indicator red or pink, and they react with bases to form water. A solution of sodium bicarbonate nicely displays several of the characteristics of basic solutions: it tastes bitter, it turns purple-cabbage indicator blue, and it reacts with acids to form water. The last property, listed for both acid and base, the ability to react with each other, is really the defining property because acid-base reactions, like redox reactions, occur in tandem: one substance acts as an acid and one substance acts as a base. Acid neutralizes base and base neutralizes acid.

People have understood the notion of neutralization—the negation of the properties of one substance by another—since the earliest pharmacopeias prescribed ground-up, toasted egg shell as a comfort for stomach problems. Egg shell is principally calcium carbonate, the same material that makes up chalk, and this material, being a base, can cancel excess stomach acid, the acid that is produced naturally in the stomach to digest food. Basic materials are also called *alkaline* materials, partly from tradition—the word *alkali* is Arabic in origin, as are *alchemy, alcohol, algebra, alcove,* and *algorithm*—and partly to avoid confusion: calling something a basic material sounds too much like calling it a fundamental material rather than something that reacts with an acid.

Many examples of acid-base reactions can be found in cooking, such as the soda–sour cream reaction in *Little Men*. In addition, the unfortunate result of acid rain is that the acid in the rain reacts with the carbonates found in limestone and marble, which causes the deterioration of statues, some of which had managed to survive without corrosion for thousands of years before the advent of the industrial age. (But before one completely condemns the industrial age, it should be remembered that the germs of Black Death, smallpox, and syphilis also managed to survive for thousands of years before modern technology brought them to bay.) This ability of acid rain to dissolve marble brings up another property common to all acids and bases: they are corrosive.

The corrosive property is addressed in the quotation from the story *Daddy Long Legs* given in the introduction and repeated here for convenience.

> *I must go to the laboratory and look into a little matter of acids and salts and alkalis. I've burned a hole as big as a plate in the front of my chemistry apron, with hydrochloric acid. If the theory worked, I ought to be able to neutralize that hole with good strong ammonia, oughtn't I?*
>
> —Jean Webster, *Daddy-Long-Legs*, 1912

The joke is that base neutralizes acid, so the chemistry student supposes that an application of base will reverse the effects of the acid and restore the destroyed material of the apron. It is interesting to note that readers in the early 1900s were supposed to understand the joke without explanation—and that first-year chemistry students in the early 1900s were encouraged to handle chemicals sufficiently active to eat a hole in a chemistry apron. But the concept of acidity is not exactly unfamiliar in our twenty-first century. We speak of an acid tongue, an acid smile, an acid voice, or acid criticism. Acid solutions are part of our kitchens: vinegar is an acetic acid solution, lemon juice is a citric acid solution, and carbonated sodas contain carbonic acid. Basic solutions and substances are also part of our common experience: ammonia is a basic solution, as is the lye used to unclog drains. Many times we use the older terms—alkali or alkaline—to describe basic materials, such as "alkaline batteries" or "adjusting the alkalinity of the swimming pool." Alkalosis, the dangerous condition of blood made too basic by the lack of carbon dioxide, can be caused by hyperventilation.

In fact, the entire human body can function well only within a very narrow range of alkalinity and acidity. But before you toss out the hot sauce for health's sake, be assured that the body has devised a method for maintaining the blood at the alkalinity level necessary for proper functioning even if vinegar is used in the salad dressing or seltzer water is consumed. The blood is *buffered*: it is a solution of significant quantities of both weak acids and weak bases so that small additions of acid or base will not cause a notable shift in alkalinity. A good analogy for a buffer might be a punching bag: an unanchored opponent might be sent reeling with a right jab, but the anchored punching bag absorbs the blow and comes back for more. A buffered solution can absorb the blows of added acid or base and bounce back to nearly its original condition. The situation can be demonstrated with another bodily fluid that is buffered: milk.

Take two of the small clear containers suggested in the "Shopping List and Solutions" and add a half cup (120 milliliters) of whole milk to each. It is necessary to use *whole* milk because cows care little about calories and provide buffers that work with the intact, not skimmed, substance. Take two more of the small clear containers and add a half cup (120 milliliters) of water to these. To all four samples add several drops of the phenol red indicator from the swimming pool test kit suggested in the "Shopping List and Solutions." The purple-cabbage indicator does not work as well in this demonstration because you have to add too much to see a vivid color and the excess liquid dilutes the milk and compromises the buffering action. Adding several drops of swimming-pool-test-kit pH indicator should give both the milk and the water a pale yellow tinge.

Now add a drop of baking soda solution (the liquid over the undissolved solid) to one of the samples of water and one of the samples of milk. If you have managed to obtain some eyedroppers, these work best, but if not, you can dip a straw in the solution and cover the end with your finger to capture a drop. Stir the solutions. The baking soda–water solution should turn a shocking pink, but the milk should stay pale yellow.

The best comparison is made by putting the containers on a white sheet of paper and looking directly down into the container. There is a considerable difference in the color of the water-indicator solution with and without baking soda. With and without baking soda, the indicator-milk solutions look the same. The milk has buffered, or moderated, the effect of the sodium bicarbonate.

In the blood, it is carbon dioxide that provides the buffer. Carbonated soda is a solution of carbonic acid formed when water is saturated with carbon dioxide, and in the purple-cabbage demonstration we saw that carbonated soda is slightly acidic. Our blood is infused with carbon dioxide, the end product of metabolism, and this carbon dioxide forms carbonic acid in the blood. As we saw in our purple-cabbage demonstration, sodium bicarbonate, baking soda, is a base. Bicarbonate, formed by removing just one hydrogen in carbonic acid, is produced by the kidneys to act as a base in the blood. So carbon dioxide is used to form both blood acid and blood base. When the blood is too acidic, the bicarbonate base sops up, neutralizes, the excess acid. When the blood is too alkaline, carbonic acid neutralizes the excess base. But as can be demonstrated by pouring baking soda solution directly into the indicator-milk solution, buffers do not have an unlimited ability to absorb acid or base: the milk will soon assume a pink shade of its own. In blood, if the balance tips too far one way or the other, the conditions of alkalosis or acidosis can occur, which are very serious medical conditions.

The blood is not the only system in the body that has to have a fairly stable acid-base environment. Buffered aspirin is aspirin that has been altered so that it will not be too acidic and upsetting to the stomach. Many soaps and shampoos are specifically blended so that they do not differ significantly from the acid quality of the skin or hair. If there were a difference in acid quality, they could cause an acid-base reaction when applied. An acid-base reaction on the skin or in the eyes could cause damage to sensitive cells, which is why you are careful to use gloves and eye protection while doing these demonstrations.

These acid- or alkaline-balanced products may be advertised as "pH balanced." But what does pH mean? We can get a clue from revisiting the pool test kit. The phenol red indicator is also labeled "pH indicator": pH is a measure of how acidic or basic a solution is. It is a measure of the acid quality of a solution.

Because there are many pairs of substances in which one substance ameliorates, or neutralizes, the effect of another, there are many substances that can be classified as acids and bases. However, we will focus on the most common group here: substances that produce the polyatomic ions of *hydronium* and *hydroxide*, the two principles of acid and base in water. *Poly* in *polyatomic* simply means more than one, as in *polytheist* or *polygamy*. Polyatomic ions are ions made up of more than one atom. The hydronium ion is created when a hydrogen ion from an acid combines with water. The resulting polyatomic ion has a positive charge: $H_3O^+$. The hydroxide ion forms when a base removes a hydrogen ion from water. The resulting polyatomic ion has a negative charge: $OH^-$. The pH of a solution is a measure of how many hydronium ions there are in a sample of the solution, acidic solutions having more hydronium ions and alkaline solutions having fewer. Note that if a hydroxide and a hydronium ion were to come together, the extra hydrogen on the hydronium ion could combine with the hydroxide ion; thus, the positive charge would cancel the negative charge, and the result would be two molecules of water. Acid and base neutralize each other.

$$H_3O^+ + OH^- \rightarrow H_2O + H_2O$$

Or, in a more compact notation,

$$H_3O^+ + OH^- \rightarrow 2\ H_2O$$

In general, the pH scale ranges from 0 to 14, a pH of 7 being neutral. But pH is not a linear measure like a ruler or a thermometer; a pH scale is another logarithmic scale like the decibel scale. It's Godzilla versus Mighty Mouse situation such as we encountered when displaying the relative abundances of the elements. To picture both acidic and basic solutions on the same chart, we have to choose a nonlinear scale. On the pH scale, a solution registering 3 is ten times as alkaline as a solution registering 2; a solution registering 4 is one hundred times as alkaline as a solution registering 2. More simply put: the greater the pH, the more basic and less acidic the solution.

But why is a pH indicator found in a swimming pool test kit? Why is it important to know how acidic or alkaline a swimming pool is? Let's plunge into that topic.

## For Example: Swimming Pool pH

Swimming pool chemistry is fairly convoluted and we will be probing more as we go along. But before going off the deep end, we will consider pH here and how it affects swimming pool chemistry.

One reason pH control is important, of course, is that people don't want to swim in solutions as acidic as vinegar or as alkaline as ammonia. For another reason, pools would be big stagnant ponds without all the pumps and filters, excellent breeders of algae, bacteria, and mosquitoes. To counteract all the natural growth, chlorine is often added to the water in the form of sodium hypochlorite, the same substance that is the active ingredient of household bleach. Sodium hypochlorite is a weak base, as bicarbonate is a weak base. However, the solution to the natural-growth problem is not as straightforward as dumping in a little bleach. The chlorine levels must stay high enough to inhibit bacterial growth without irritating the eyes and noses of the swimmers. Chlorine levels must also remain fairly constant while being exposed to sunlight, variations in temperature, and metal, plastic, and ceramic surfaces.

The best antibacterial action is derived from the bleach when the pool water is acidic, but acidic water is not pleasant to swim in and can be corrosive to the metal fittings of plumbing, filters, and pumps. Similarly, a swimming pool solution that is too alkaline can also be corrosive. So a balance must be struck. Anyone who has ever owned a pool or has had experience with pool maintenance is aware that this can be a daily chore during times of peak pool usage. It would be an hourly chore were steps not taken to buffer the pool. In swimming pools, the buffering action is achieved by having sufficient quantities of both bleach and acid. In pool maintenance manuals, "sufficient quantities" is termed "total alkalinity" and is a quantity that must be measured and regulated along with the pH and chlorine content of the pool. The acid content of the pool is adjusted by adding swimming pool acid and the alkalinity is adjusted by adding the prescribed base. However, if too much base is added, another problem results: *scale*.

Scale is a rock-hard crust that can form in pipes and pots that are used with hard water. Before the general availability of household water softeners, scale was a much more common experience. Insoluble scale forms from calcium ions when carbonate ion is present. This fact highlights, once more, the versatility of carbonates. We have seen carbon dioxide form carbonates and hence carbonic acid in water; we have used sodium bicarbonate (baking soda) as a base; and finally we have pointed out that carbonates can make fairly insoluble solids. These many talents of the carbonate ion make baking soda good for more than cooking. Baking soda makes an excellent deodorizer because it can react with both acidic and basic smelly compounds and can form nonvolatile, and hence nonsmelly, compounds with many more. A lot of chemistry in a little box!

Some calcium and other minerals are found naturally in any water that has percolated through the ground. So if the swimming pool is allowed to become too alkaline, it will grow cloudy, at best, with suspended particles of insoluble salts or encrusted with growths of scale at worst. This appearance of a solid in a liquid solution is called *precipitation*, just as rain is called *precipitation*, because of the propensity of the solids to come out of the solution just as rain comes out of the sky.

In the next chapter, we look at some other solid-forming solutions, as well as some whys, wherefores, and useful applications of these crusty precipitation reactions.

# Demonstration 5: Blue Blob, Black Ink

*Dr. Archie watched her contemplatively, as if she were a beaker full of chemicals working.*
—Willa Cather, *The Song of the Lark*, ca. 1915

## Blue Blob

Adjust your safety glasses comfortably over your eyes, and then add a tablespoon (15 milliliters) of baking soda to a half cup of water (120 milliliters) and stir. Allow the undissolved material to settle to the bottom (this should take about a minute), and then carefully pour off the clear liquid into another glass, leaving the solids. Pour a tablespoon (15 milliliters) of the copper sulfate solution described in the "Shopping List and Solutions" into the decanted baking soda solution. Beautiful blue flakes should immediately form and settle out of solution. You should also see some bubbles, and the *supernatant*, the liquid over the blue blob precipitate, should be a pastel blue. The bubbles are from excess acid in the copper solution reacting with the bicarbonate ion.

The blue blob reaction occurs because the water decanted from the baking soda contains dissolved sodium and bicarbonate. A small amount of the bicarbonate loses a hydrogen to water and becomes the carbonate ion. The copper solution has copper ions and sulfate ions. When the copper ion finds the carbonate ion, it is love at first sight. These two ions combine and hold on tight. In fact, they hold on so tightly they form a solid and fall out of solution. They precipitate.

## Black Ink

With safety glasses securely in place, add a teaspoon (5 milliliters) of hydrogen peroxide to a quarter cup (60 milliliters) of iron acetate solution (see the "Shopping List and Solutions"). The solution should turn a nice reddish brown as the ferrous ion becomes the ferric ion. Slowly add a half cup (120 milliliters) of cold, brewed brown tea. You should get a black, mushy precipitate that will eventually settle to the bottom. Historically, this material was used as black ink, and this precipitation reaction was used to produce it.

# CHAPTER 5

## Chemical Partners: Who Does What to Whom

> *(T)his last misdeed had a decisive effect upon him; it rushed across the chaos of his intellect and dissipated it, set the light on one side and the dark clouds on the other, and acted upon his soul . . . as certain chemical reagents act upon a turbid mixture, by precipitating one element and producing a clear solution of the other.*
>
> —Victor Hugo, *Les Misérables*, ca. 1860

To a chemist, saying a solution is "clear" is not saying the solution is without color. Saying a solution is clear means it has no suspended solids. A precipitation reaction is one in which two clear solutions mix to form a solid in solution. Why should two clear solutions decide to form a solid? Consider the following analogous social situation. Two celebrants may come to a party together because it is convenient to share a ride. In the course of an evening, however, they separate and come in contact with other people. If, across the crowded room, they find someone with whom they form a special bond, they may leave the party with this person and, with luck, stay bonded. If, across the crowded beaker, an ion should find another ion with which it forms a strong bond, a precipitate will result. When two ions join to form a solid, the solid is called a *salt*.

The social analogy can be extended—unfortunately. Even strongly bonded chemical species can be enticed to change partners temporarily. Put your safety glasses back on again for a minute. Dissolve a teaspoon (5 milliliters) of alum in a half cup (120 milliliters) of water. Add less than a quarter teaspoon (1 milliliter) of ammonia and stir. You should see a thick white precipitate. Add four drops of aquarium pH-lowering solution and stir. Keep adding drops until the white precipitate dissolves and you have a clear solution again. Ammonia is a base, and the aquarium pH-lowering solution is an acid. (Recall that a lower pH means a more acidic solution.) Whether or not a precipitate will form will depend on the salt's environment.

The social analogy can be extended again. Dissolution is the opposite of precipitation, and in the complete absence of water, all unions of salts are solid. Water, however, can be a home wrecker. When a salt is wet, the molecules of water can insinuate themselves between ions. Table salt, sodium chloride, is solid enough in the shaker, but when it is wet, the molecules of water work their way between the chloride ion and sodium ion and keep them apart. Why doesn't water come between the carbonate ion and the copper ion in the insoluble copper carbonate of the earlier demonstration? It boils down to a trade-off: if the divorce is more costly than any possible benefit, the union stays solid. If not, let the proceedings begin.

To understand the factors involved in the cost/benefit analysis for salt dissolution, we first need to revisit an idea introduced earlier. We said that atoms like to gain or lose electrons if doing so brings them closer to their ideal of a filled shell, but we also said that when nonmetals combine, they tend to share, rather than appropriate, electrons to fill their needs. As it turns out, some elements have a stronger attraction for electrons than others. The ability to attract electrons has been termed *electronegativity*, and the degree of electronegativity depends on the positions of the elements on the periodic table.

Of all the elements on the periodic table, fluorine is the most electronegative, as shown in figure 1.5.1. This means that fluorine is the most likely to command the lion's share of the electrons. In a fluorine compound, the electrons will cluster around fluorine more than any other element. On the periodic table, the closer an element is to fluorine, the more electronegative it is. Because electronegativity is a measure of the element's attraction for electrons when it is in combination with another element, electronegativity isn't a concept for the noble gases, helium (He) through radon (Rn). This family doesn't form compounds under normal conditions.

Electronegativity →

Electronegativity ↑

| | | | | | | | | | | | | | | | | | |
|---|---|---|---|---|---|---|---|---|---|---|---|---|---|---|---|---|---|
| H | | | | | | | | | | | | | | | | | He |
| Li | Be | | | | | | | | | | | B | C | N | O | **F** | Ne |
| Na | Mg | | | | | | | | | | | Al | Si | P | S | Cl | Ar |
| K | Ca | Sc | Ti | V | Cr | Mn | Fe | Co | Ni | Cu | Zn | Ga | Ge | As | Se | Br | Kr |
| Rb | Sr | Y | Zr | Nb | Mo | Tc | Ru | Rh | Pd | Ag | Cd | In | Sn | Sb | Te | I | Xe |
| Cs | Ba | La* | Hf | Ta | W | Re | Os | Ir | Pt | Au | Hg | Tl | Pb | Bi | Po | At | Rn |
| Fr | Ra | Ac† | Rf | Db | Sg | Bh | Hs | Mt | • | • | • | | | | | | |

Figure 1.5.1. A periodic table showing the trends in electronegativity. Elements tend to become more electronegative toward the right and to the top of the periodic table. Fluorine is the most electronegative element. The noble gases are not assigned an electronegativity because they do not normally enter into chemical reactions.

Isn't that convenient? Not only does the periodic table tell us the number of protons in the nucleus, the number of electrons in a neutral atom of the element, and something of an element's reactivity from the family it belongs to, but the periodic table can also be a device for predicting relative electronegativity. The explanation for this correspondence comes from understanding atomic structure: the elements toward the top of the periodic table don't have very many filled shells, so a new electron can feel more of the attraction of the positive nucleus without other electrons blocking the way. Fluorine enjoys its status as the most electronegative element on the periodic table primarily because fluorine has only one filled inner shell, and second, of all the elements with a partially filled second shell, it has the greatest nuclear charge: nine protons. The only second-row element with a greater nuclear charge is neon, and it has all the electrons it wants. Then what about hydrogen? It has *no* filled shells, and an electron added to hydrogen would be fully exposed to the positively charged nucleus. Why doesn't hydrogen have a large electronegativity? Because even on the atomic level, it turns out, size matters.

Generally speaking, atoms tend to get bigger as you move down the periodic table because the elements have more electrons and these electrons are filling up shells. Hydrogen isn't as electronegative as fluorine because it hasn't acquired any filled shells and it is too small to comfortably support a large negative charge. The difference between a fluoride ion carrying a negative charge and a hydrogen ion carrying a negative charge is like the difference between someone carrying a twenty-pound sack of dog food or a twenty-pound cannonball. The dog food is easier to carry because the weight is spread out over a larger area. The cannonball is harder to carry because the weight is concentrated at one point. When the hydrogen atom carries a negative charge, all that charge is concentrated into a very small volume. When all the various factors weigh in, hydrogen has an electronegativity comparable to that of phosphorus, as shown in figure 1.5.2.

Electronegativity →

Electronegativity ↑

| | | | | | | | | | | | | | | | | | He |
|---|---|---|---|---|---|---|---|---|---|---|---|---|---|---|---|---|---|
| Li | Be | | | | | | | | | | | B | C | N | O | F | Ne |
| Na | Mg | | | | | | | | | | | Al | Si | P<br>H | S | Cl | Ar |
| K | Ca | Sc | Ti | V | Cr | Mn | Fe | Co | Ni | Cu | Zn | Ga | Ge | As | Se | Br | Kr |
| Rb | Sr | Y | Zr | Nb | Mo | Tc | Ru | Rh | Pd | Ag | Cd | In | Sn | Sb | Te | I | Xe |
| Cs | Ba | La* | Hf | Ta | W | Re | Os | Ir | Pt | Au | Hg | Tl | Pb | Bi | Po | At | Rn |
| Fr | Ra | Act | Rf | Db | Sg | Bh | Hs | Mt | • | • | • | | | | | | |

Figure 1.5.2. Hydrogen has an electronegativity comparable to that of phosphorous.

So what is the result of all these attractions and electron swappings among the elements in a compound? Something called *molecular polarity*.

Polarity is a familiar concept. Batteries have positive poles and negative poles. Magnets have north poles and south poles. When the electrons preferentially sit on one end of a molecule because that is where the electronegative elements are, the molecule has a positive pole and a negative pole; thus, it has polarity. A molecule with a more positive end and a more negative end is said to have a *dipole* as it has "two poles."

Polar molecules behave somewhat as bar magnets do: the positive end of one polar molecule lines up with the negative end of another, as the north pole of one bar magnet will line up with the south pole of another. Water ($H_2O$), shaped in a **V** with the oxygen in the middle and the hydrogens at the tips of the **V**, is a polar molecule. Oxygen is more electronegative than hydrogen, so the electrons gravitate toward the oxygen end and away from the hydrogen end. The negatively charged oxygen ends of several water molecules can be attracted to a positively charged ion, building a cage around it and keeping it separated from its negatively charged partner. Similarly, the positively charged hydrogen ends of several water molecules can align themselves with negatively charged ions, building a similar cage around them. This situation is shown schematically in figure 1.5.3.

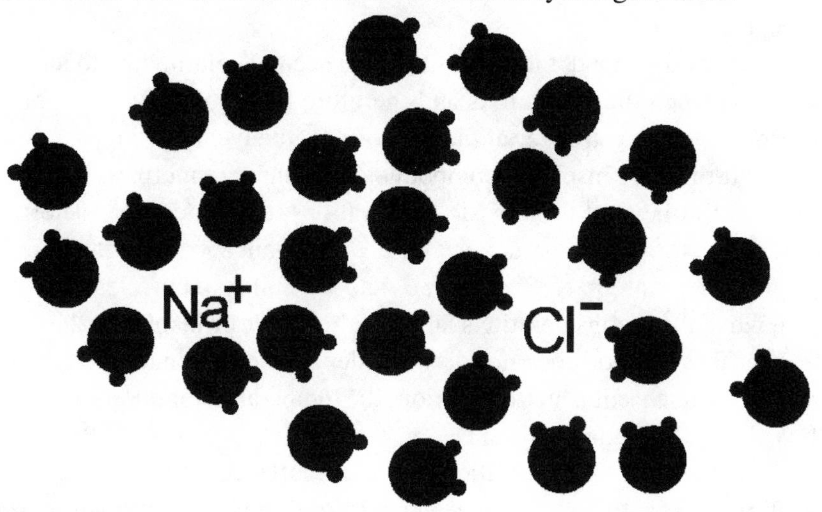


Figure 1.5.3. Water molecules separating a sodium ion and a chloride ion by forming solvent cages around the ions.

Water molecules like to build cages around ions because this is a more relaxing situation for them. Just as it requires energy to keep magnets apart that are attracted to each other, it takes energy to hold apart molecules that are attracted to each other or to an ion. Allowing water molecules to form cages around ions allows them to relax and release energy.

However, energy is also needed to break the salt apart, so there is the trade-off. If it costs more to break the salt apart than the profits gained by allowing the water molecules to relax into cages, then the salt will precipitate out of solution or not dissolve in the first place. If it costs less to break the salt apart than the profits gained by allowing the water molecules to build cages, then the salt will stay in solution or dissolve.

By revisiting the periodic table, we can conclude that the metals in the first column (lithium [Li], sodium [Na], potassium [K], and so on) like to form positive ions with a positive-one charge because losing one electron will give them a nicely filled outermost shell. Salts derived from these positive ions are almost always soluble because the attraction of one positive charge is relatively weak. Compare the bar magnets suggested in the "Shopping List and Solutions" with a refrigerator magnet. The positive ions formed from the elements in column one would have a relatively weak attraction for negative ions, more like the refrigerator magnet than the bar magnet.

On the other hand, the elements of the second column like to lose two electrons when they form ions and therefore have a positive-two charge. These ions have a stronger attraction for negatively charged ions and will tend to form more insoluble compounds than their counterparts in the first column. But beyond these basic observations, the factors that determine which salts form more permanent unions are again about as convoluted as the human counterpart. Size, shape, and flexibility play roles, too. And even when these considerations label a salt soluble or insoluble, the actual amount that does or does not dissolve depends on the nature of the solvent, how much salt is in the solution, the temperature of the solution, and the pH of the solution.

A convincing demonstration of the importance of the nature of the solution is to try to dissolve a teaspoon (5 milliliters) of table salt in a cup (240 milliliters) of water versus the same quantity in salad oil or mineral oil. Table salt in water will dissolve readily. Table salt in oil settles to the bottom, undissolved. Try stirring both before making a final determination. Stirring helps salts dissolve because the dissolution process starts at the solid's surface and stirring can help expose more surfaces to the solution. However, no amount of stirring will force all the table salt into an oily solution.

The quantity dependence of solubility can be demonstrated with baking soda and water. If a pinch of baking soda is added to a clear glass (not plastic) holding about two cups of water, the pinch should dissolve. In fact, up to ten or fifteen healthy pinches of baking soda should dissolve. At some point, however, the added pinches start turning the solution cloudy, and eventually the added pinches settle to the bottom, even with stirring. To demonstrate the effect of temperature on solubility, the solution can be placed in the microwave for thirty seconds, just enough time to warm it to the temperature of hot but drinkable soup. The resulting warm solution should be clear again, and all the baking soda on the bottom of the glass should dissolve with stirring.

The influence of other materials in solution can be shown with chalk. Compare chalk, which is calcium carbonate, with baking soda, which is the compound formed from sodium, hydrogen, and carbonate. When a teaspoon (5 milliliters) of baking soda is added to a cup (240 milliliters) of warm water, it dissolves. The same amount of powdered chalk would sink to the bottom like a rock. However, chalk will dissolve when placed in vinegar. Because vinegar is an acid and chalk is reacting with the acid, the chalk, by definition, is a base, but a base that will not dissolve in neutral solutions such as plain water. In vinegar, an acid, the carbonate base is neutralized and calcium goes into solution.

So with all of these variables—the nature of the salt, nature of the solution, amount of salt, and temperature of the solution—it seems that precipitation and solubility are fairly complicated. Things don't just immediately fall into place. We will nonetheless be revisiting these concepts in many different guises as we proceed because precipitation and solubility factor into a lot of chemical situations. For instance, seashells form when calcium excreted from a sea creature mixes with the carbonate in the water to form calcium carbonate. Eons of seashells collecting on the bottom of the ocean account for the composition of the chalk and limestone that finds its way to our blackboards and sidewalks.

Precipitation is useful in hazardous waste treatment. If harmful metals can be precipitated out of solution, they can be segregated as solids that take up less space than liquid waste and are easier to transport and store. Other, troublesome, precipitation reactions occur in an unexpected place: the human body. Gout is a condition caused by uric acid precipitates accumulating in the joints. Gallstones and kidney stones are precipitates. These pathological precipitates, however, are not the norm. The body is a pretty soupy mixture because the essence of life is the ability to change in response to the environment, so most bodily substances must be able and ready to move. This mobility can be accomplished with water because water is so very good at keeping substances in solution. The body can last much longer without food than without water, and without water, the body and mind become lethargic. Many conditions, as well as moods, can be positively influenced, if not completely corrected, by increasing the daily intake of water. Many times, when it comes to health, water is the solution.

Another common precipitation product is soap scum. Soap scum generally floats on water rather than raining out, so it is normally not thought of as a precipitate—but that is a matter of semantics. Scum can form in bathtubs when soap mixes with hard water; that is, water that contains calcium and magnesium ions. Water that contains calcium and magnesium ions is called "hard" because these ions tend to form insoluble precipitates (that is, little rocks) in the water.

## For Example: Hard Water, Soft Water

The good thing about water is that so many things dissolve in it. The bad thing about water is that so many things dissolve in it. Pure water is a poor conductor of electricity, but don't go standing in a full bathtub to change a lightbulb because it is very difficult to obtain pure water. Water with salts dissolved in it is a good conductor of electricity. The ability of water to dissolve things accounts for the fact that all groundwater contains dissolved ions, including calcium and magnesium salts.

As dropping a piece of chalk in water confirms, calcium salts can be pretty insoluble, and while the buildup of calcium salts on tea kettles can be annoying and scum can be frustrating, the buildup of crusty deposits of scale in pipes and swimming pools can precipitate serious problems. At one time, negatively charged phosphate ions were added to detergents so the calcium and other positively charged ions would preferentially form salts with the phosphates and leave the detergent free to perform its function. But phosphates are quite efficient fertilizers. So waste streams from laundry caused waste water conduits to become clogged with unusually dense growth, which is why phosphate use had to be restricted. Nowadays scum formation is not so much of a problem because of the widespread use of water softeners. Water softeners remove calcium and other precipitate-forming ions from the water before they go through the household pipes.

The principles at play in water softeners are the same as those in precipitation reactions. Water is run over a dense material saturated with sodium ions like a gigantic network of salts. The sodium salts are soluble, whereas the calcium analogs of the same salts are not. So the calcium replaces the sodium and becomes trapped on or in the solid. The water, now enriched in sodium instead of calcium, can be used without worry that it will precipitate soap or form the crusty deposits of scale. In the regeneration cycle of the water softener, the solid support is backwashed with water saturated with so much sodium salt that the sodium eventually replaces the calcium on the solid, and the water softener is ready to start softening again.

But a material doesn't have to be a salt to dissolve in water. Sugar, decidedly not a salt, dissolves in water, too. Which brings up the question: What makes sugar so different from salt? They are both white crystalline compounds that dissolve in water and are difficult to tell apart without tasting. But you could survive for a least a while on a diet of sugar water; on saltwater, you would not survive. What is the difference? It's all in the bond. The chemical bond.

## Demonstration 6: Bond. Chemical Bond.

*Only the chemist can tell, and not always the chemist,*
*What will result from compounding*
*Fluids or solids.*
*And who can tell*
*How men and women will interact*
*On each other, or what children will result?*
*There were Benjamin Pantier and his wife,*
*Good in themselves, but evil toward each other:*
*He oxygen, she hydrogen,*
*Their son, a devastating fire.*

—Edgar Lee Masters, *Spoon River Anthology,* ca 1916

*I fear . . . the chemical action which would be set up in my soul by a false homage to a symbol.*

—James Joyce, *Portrait of the Artist as a Young Man,* ca. 1915

Take one of the clear plastic bottles suggested in the "Shopping List and Solutions," put on your safety glasses, and pour in about two inches (5 centimeters) of mineral oil. Add about two inches (5 centimeters) of copper sulfate solution. Put the cap on the bottle and shake it. The mixture should separate into two layers with the copper solution layer remaining bright blue and the mineral oil layer remaining clear. Add a quarter teaspoon (1 milliliter) of tincture of iodine. Put the cap on and shake.

When you stop shaking, the two layers should separate again, but this time the mineral oil layer should appear violet. Tincture of iodine contains both $I^-$ and $I_2$ but mostly a complex of the two, which is brown. The copper ions in the copper sulfate oxidize $I^-$ to the molecular form $I_2$, which ruins the complex. The molecule $I_2$ is violet and soluble in oils such as mineral oil.[1]

Now add a teaspoon (5 milliliters) of household ammonia. Put the cap on and shake. This time when the layers separate, the copper layer should be a dark royal blue because the copper is now forming strong partnerships with ammonia and it is no longer available to the iodine. Some iodine reverts to $I^-$, the complex re-forms, and the mineral oil clears up. You may need to add several portions of ammonia, but eventually the mineral oil clears.

The molecule $I_2$ dissolves in the mineral oil, but the ion $I^-$ does not. The ion $I^-$ dissolves in water but not in mineral oil. Water, a polar molecule, is attracted to the ion $I^-$, builds a cage around it, and pulls $I^-$ into the water solution (otherwise known as *aqueous* solution). The molecule $I_2$ has no polar end to attract water molecules, so the water molecules are more attracted to each other and squeeze $I_2$ out of solution. The mineral oil accepts $I_2$ because mineral oil is a nonpolar molecule and has no cage-building propensities. Earlier we said that salts were formed when positive ions are attracted to negative ions and that these materials form a bond. Now we need to add that this type of bond is, specifically, an *ionic bond.* This special designation is necessary because not all materials are salts and not all chemical bonds are ionic. The bonding that occurs in non-salty materials, such as $I_2$, is called *covalent.* The differences between the types of bonding and the materials they produce deserve a chapter of their own, such as the one that follows.

# CHAPTER 6
## The Tie That Binds, the Chemicals That Bond

*Close relation in a place like London is a personal mystery as profound as chemical affinity. Thousands pass, and one separates himself from the mass to attach himself to another, and so make, little by little, a group.*

—Henry Adams, *The Education of Henry Adams,* ca. 1920

*Polarity, or action and reaction, we meet in every part of nature; in darkness and light; in heat and cold; in the ebb and flow of waters; in male and female; in the inspiration and expiration of plants and animals; in the systole and diastole of the heart; in the undulations of fluids, and of sound; in the centrifugal and centripetal gravity; in electricity, galvanism and chemical affinity.*

—Ralph Waldo Emerson, *Essays,* ca. 1840

The illustration for the demonstration with which we began this discussion was a martini glass with two olives. But unlike the toothpick holding the olives together, a bond is not a material entity. A bond is a force, in the same sense that magnetism and gravity are forces. The origin of the force in a chemical bond is electrical, but electrical interactions that are complicated by attractions and repulsions, the presence of many electrons and protons, and charges that are moving in complicated patterns. Trying to categorize substances by type of bond is somewhat like a cosmetician trying to categorize humans by skin tone. Certainly there are dark-skin people and light-skin people, but there are also many variations within the range of human skin. In the same way, we can say there are three basic categories of chemical bonds—ionic, covalent, and metallic—but these are really just variations on a theme: nuclei are bonded together when there is a mutual attraction, and the basis for this mutual attraction is electrical. All bonds have a bit of each quality—covalent, ionic, and metallic—to a greater or lesser degree. Such fuzziness is not immediately useful, however, so for now we will proceed as though it were possible unequivocally to place each molecule in one category or another and use the categorization to assign properties to the different materials.

To begin with, we have an excellent tool for predicting just exactly what kind of ion an element is most likely to form—and that is the chemist's closest friend, the periodic table—which is shown here again for easy reference.

| H | | | | | | | | | | | | | | | | | He |
|---|---|---|---|---|---|---|---|---|---|---|---|---|---|---|---|---|---|
| Li | Be | | | | | | | | | | | B | C | N | O | F | Ne |
| Na | Mg | | | | | | | | | | | Al | Si | P | S | Cl | Ar |
| K | Ca | Sc | Ti | V | Cr | Mn | Fe | Co | Ni | Cu | Zn | Ga | Ge | As | Se | Br | Kr |
| Rb | Sr | Y | Zr | Nb | Mo | Tc | Ru | Rh | Pd | Ag | Cd | In | Sn | Sb | Te | I | Xe |
| Cs | Ba | La* | Hf | Ta | W | Re | Os | Ir | Pt | Au | Hg | Tl | Pb | Bi | Po | At | Rn |
| Fr | Ra | Ac† | Rf | Db | Sg | Bh | Hs | Mt | • | • | • | | | | | | |

| *Ce | Pr | Nd | Pm | Sm | Eu | Gd | Tb | Dy | Ho | Er | Tm | Yb | Lu |
|---|---|---|---|---|---|---|---|---|---|---|---|---|---|
| †Th | Pa | U | Np | Pu | Am | Cm | Bk | Cf | Es | Fm | Md | No | Lr |

Figure 1.6.1. The periodic table of the elements.

As we've pointed out, the periodic table can be broken down into three major parts: the metals, the nonmetals, and the metalloids. The metals are to the left of the zigzag staircase that starts between B (boron) and Al (aluminum) and ends between Po (polonium) and At (astatine). The nonmetals are to the right of the staircase. The metalloids are the elements immediately above and below the staircase. But once again, there is a range of properties and no exact demarcation. Nature doesn't care what we choose to call metal and what we choose to call nonmetal. However, some elements (such as copper, Cu, three elements to the left of the staircase) are clearly in the metal block. They have behaviors that are clearly metal-like (copper solid is shiny and bendable and is a great conductor of electricity). The behaviors of other elements in the nonmetal block are clearly nonmetal-like: nitrogen gas and oxygen gas are such poor conductors of electricity that current flow can be stopped by simply separating wires in air. So the division of the periodic table into metals and nonmetals can be a useful organizing principle.

The nonmetals to the right of the staircase usually like to form negative ions when they form ions. Metals to the left of the staircase usually like to form positive ions when forming ions. Opposite charges attract. When ions of opposite charges are in the same vicinity, they are attracted to each other. If they form a bond based on this attraction between ions, the bond is called an ionic bond. For instance, the bond between sodium and chlorine in table salt, sodium chloride, is a very good example of what is considered an ionic bond. The ionic bond originates from the attractions between positive and negative ions, as the name implies, and ionic bonds are so wonderfully predictable and explainable that one almost wishes that all bonds could be simple ionic bonds. But then we wouldn't have the virtually infinite variety of materials that are necessary to the world we know. Variety is always an interesting, but complicating, factor.

As we noted previously, ions form because atoms tend to gain or lose electrons until they have a filled valence shell. For historical reasons, a filled shell is called an *octet*, despite the fact that not all filled shells consist of eight electrons. We speak of ions being formed when atoms gain or lose electrons to fill their octets, but there are ways atoms can fill their octet that do not involve abandoning or stealing electrons. Octets can be completed by sharing electrons, too.

Electron sharing results in the type of bonding that usually occurs between two nonmetals (that is, two elements that are both on the far right of the periodic table). This type of bonding is termed *covalent* bonding. As we've seen, the valence electrons are those electrons in the outermost shell of an atom. For instance, carbon, in the fourth position on the second row, has six total electrons (being the sixth element listed on the periodic table), but only four of these are valence electrons, that is, electrons in the outer shell. *Covalent* means that electrons are being shared cooperatively and act as valence electrons for both the elements in the bond.

The theoretical rationale for electron sharing is that both elements in a covalent bond have about the same attraction for electrons. Being on the same side of the periodic table and having similar electronegativity, there would be no way of deciding who gets the electrons and who gives them up. Oxygen is on the right on the periodic table, and, as we said before, hydrogen has an electronegativity similar to elements on the right of the periodic table. So the bond between oxygen and hydrogen in water is covalent. When the valence electrons from oxygen and hydrogen finish arranging themselves around the nuclei, each element has a filled octet. (Recall that hydrogen's "octet" is full with two electrons.) Two electrons fill the octet for each hydrogen, and eight electrons fill the octet for oxygen, four of which are electrons shared with the hydrogen nuclei.

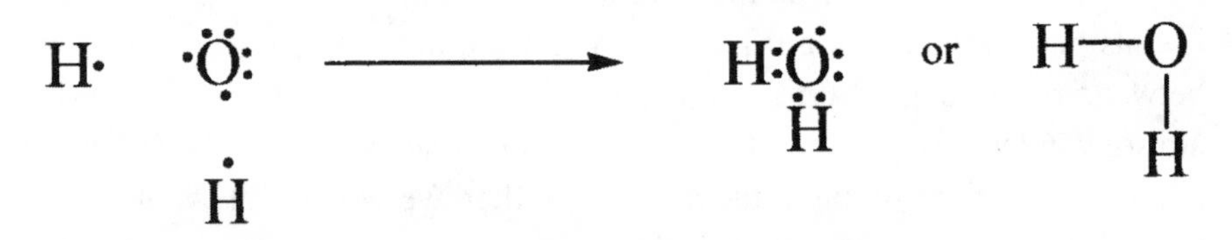


Figure 1.6.2. Covalent bonds form between hydrogen and oxygen when they fill their respective shells by sharing electrons.

There is, however, a fundamental contradiction. We have stated that opposite charges attract and like charges repel, and that electrons all have a negative charge. If this is true, then why don't collections of valence electrons explode into flying showers of electrons instead of settling into strong chemical bonds?

That is a very good question and we are glad you asked. It turns out that this question long bedeviled chemical theorists, too. They called the observed attraction between elements *chemical affinity* just to give it a name, but they did not understand the nature of the attraction. With an understanding of atomic structure, however, came a rationalization for such attractions: they can be seen as negative electrons running interference between positively charged nuclei.

Though the complete picture is a good deal more complex, a reasonable analogy can be found in magnets. If two bar magnets are brought together so that like poles align, they repulse one another. This situation is similar to bringing two bare, positively charged nuclei together. They do not bond because of the repulsion between them. If, however, a third bar magnet is placed between the original two, with the south pole aligned with the two north poles, then the magnet in the middle will attract the other two and the three will stay together. In a bond, the negative electrons between the two positively charged nuclei attract both positive nuclei and act like a glue to hold the nuclei together.

Analogies, however, are never perfect. The electrons do not occupy some specific space between the nuclei like a third bar magnet but spread themselves out into fuzzy clouds, called *molecular orbitals*, that can assume shapes nearly as diverse at the fluffy celestial formations we have compared them to. When one of these orbitals has a significant electron density between the two nuclei, the formation acts as a bonding orbital. If one of these electron clouds results in a significant electron density outside the nuclei, the electrons can lessen the bonding rather than enhance it. In this case the electron cloud, the electron orbital, is called an antibonding orbital. These two situations are depicted in figure 1.6.3. On the left-hand side, figure 1.6.3a, two atoms have come together so that there is some electron density between the nuclei. The electrons are attracted to the adjoining nuclei and a bond is formed. On the right-hand side, figure 1.6.3b, two atoms come together so that there is a lack of electron density between them. The exposed, positively charged nuclei repel each other, and no bond is formed.

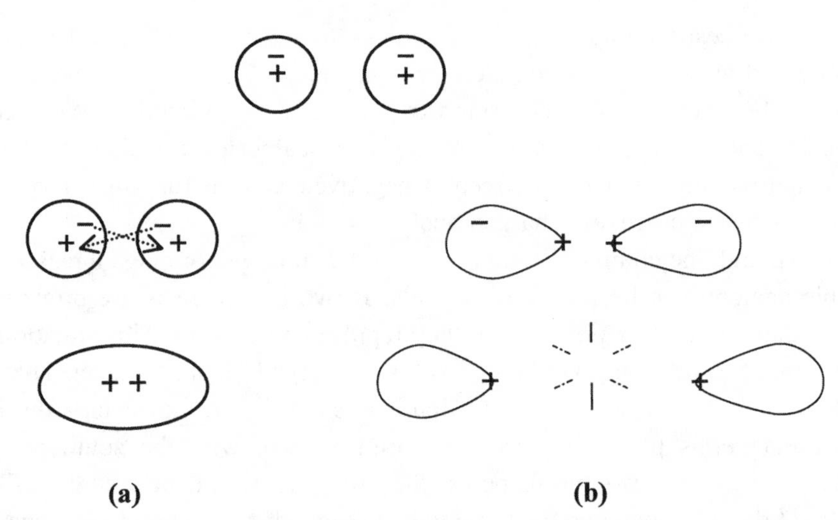


Figure 1.6.3. When two atoms come together, they may find themselves (a) with an electron cloud, between the two nuclei, or (b) outside the center of the two nuclei. In the first case, (a), there will be an attraction between the electron cloud of one atom and the nucleus of the other and a bond may form. In the second case, (b), the two positively charged nuclei are exposed to, and repelled from, each other. A bond will not form.

The antibonding orbitals are generally at a higher energy than the bonding orbitals, but if the right amount of energy impinges on the bonding orbital, perhaps in the form of light energy, electrons can be excited from one orbital to another. If electrons are excited to an antibonding orbital from a bonding orbital, the bond may break. Many substances, such as some beers, are stored in brown glass to prevent light from degrading the material inside by just such a mechanism.

As we said above, bonds aren't always purely ionic or purely covalent. The best models for ionic bonds include some electron sharing. And covalent bonds don't always share electrons completely evenly. Any difference in electronegativity between bonding nuclei means that the electrons will be attracted more to one nucleus than the other. However, a bond could be considered purely covalent if it is between two identical nuclei, as in hydrogen gas, $H_2$, or oxygen gas, $O_2$. An extreme example of equality in electron sharing is the metallic bond. In a metal, all the atoms are identical and share electrons so readily that the metal can be thought of as one big molecule.

In solids, there can be electrons in a bonding system between pairs of nuclei called the *valence band.* There can also be a gigantic orbital that extends over all of the nuclei called the *conduction band*. The conduction band can be thought of as a huge orbital. The bonding electrons in the valence band stay associated with their pair of nuclei. Electrons in the conduction band are free to move from nucleus to nucleus and travel from one end of the metal to the other. An analogy might be made to a banking network. At the local level, there are town banks with local tellers that live and work in the same place. These tellers are the valence-band electrons, keeping the foundations of the bank together. The upper-level managers can be thought of as the conduction-band electrons: their work keeps them moving from bank to bank over the entire network.

Materials that have plenty of electrons in the conduction band are, logically enough, called *conductors*, and energy supplied by a battery can cause these electrons to move in what is generally called an *electric current*. Other solids do not have electrons in a conduction band and do not conduct electricity. They are called *insulators*. Materials that are intermediate between conductors and insulators are called *semiconductors*, a fairly innocuous-sounding name for a revolutionary material.

## For Example: Semiconductors

The invention of transistors, based on semiconductors, has changed the world. What are semiconductors? As the name implies, they are materials with electrical properties between those of conductors and insulators. Semiconductors are materials that conduct some of the time.

In a semiconductor, there are normally too few electrons in the conduction band to conduct current. However, the energy gap between the conduction band and the valence band is small enough that the electrons, with a little provocation, can jump the gap and make the material conductive. This provocation can come in the form of applied voltage, heat, or light. Some solids are natural semiconductors, or what are termed *intrinsic* semiconductors. Others, however, can be made more able to jump the gap by a process called *doping*, though this has nothing to do with athletes or drugs.

Silicon is normally the first material that comes to mind when speaking of semiconductors, but silicon is, in fact, not an intrinsic semiconductor. Solid silicon must be doped to become a semiconductor. Doping is the process of adding a miniscule amount of impurity to a solid such as silicon. Two of the favorite dopants for silicon are phosphorus and aluminum because they are on either side of silicon on the periodic table. They have about the same size and electronegativity, so they can easily fit next to silicon in a solid. They change the conductive properties of silicon because they have, respectively, one more and one fewer electron than silicon.

As can be seen from the periodic table, silicon has four valence electrons available for bonding. Accordingly, it bonds with four other silicon atoms when it solidifies, giving each a shared tally of eight electrons, a nicely filled shell. If, however, an atom of phosphorus were to slip into the mix, then there would be one extra electron that didn't quite fit. This extra electron is more easily promoted to the conduction band and therefore allows the material to conduct more readily when hooked up to a battery. If an atom of aluminum, with three valence electrons, insinuates itself into the silicon solid, then there is a hole where a valence electron would normally be found. This extra hole also allows current to flow more readily because electrons motivated to move now have a place to go. Silicon doped with something that donates extra electrons is called an *n*-type semiconductor because an electron contributes a negative charge. Silicon doped with something that has fewer electrons is called a *p*-type semiconductor because it contributes a *positive* charge.

When a piece of *p*-type material is joined to a piece of *n*-type material, an arrangement called a *diode* is formed. A diode is a device that will allow current to flow in only one direction. A bit of *p*-type semiconductor attached to a bit of *n*-type semiconductor will allow current to flow only when electrons are going into the negative side, over to the positive side, and out. A battery hooked with its negative side attached to the *n*-type side of a diode acts like a river supplying water to flow over a dam. The positive end of the battery is the reservoir into which the water flows. Conversely, if the negative side of a battery is hooked up to the *p*-type material and the positive side of a battery is hooked up to the *n*-type material, then there is a tug-of-war for the electrons that ends in a stalemate and no current flows.

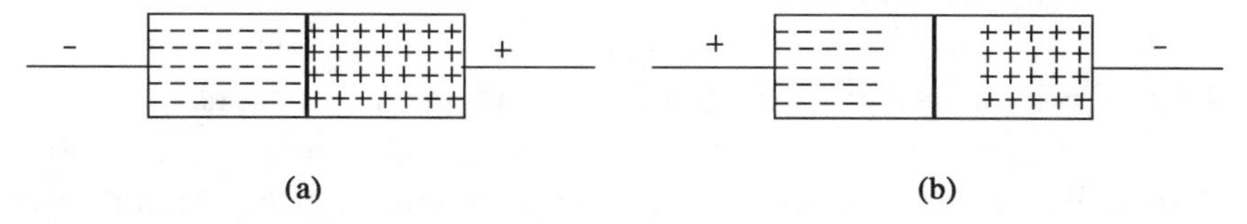


Figure 1.6.4. A diode can be constructed by joining an *n*-type semiconductor to a *p*-type semiconductor. When an electron supply is hooked up as shown in (a), current will flow through the diode. When an electron supply is hooked up as shown in (b), there is a tug-of-war for the electrons and no current flows.

This behavior becomes useful when a third bit of semiconductor material is added, making a *pnp* or *npn* sandwich. Current cannot flow in *either* direction in a *pnp* or *npn* sandwich—unless some other source injects or pulls out electrons from the middle layer of the sandwich, making *ppp*- or *nnn*-type materials, which can conduct.

The situation begs for an analogy. A *pnp* or *npn* material is like a dammed-up river. Changing the *n* in *pnp* to *p* by pulling out electrons is like opening up the dam and allowing the water to flow. The same is true for *npn* if electrons are injected into the *p* material. What injects or pulls out electrons? A battery. A battery hooked to the middle material can raise or lower the electron dam.

So what purpose does this serve? Now a small current can control a large current, switching it on and off with a little electron flow. To appreciate some of the significance of the situation, let's think historically for a moment. Europeans in the very early Middle Ages were technology poor and relied on manual labor. If they needed to grind flour, then several serfs were tethered to a grist mill and told to start pushing. The Black Death, however, decimated the ranks of the workers, so the remaining people found themselves becoming more creative with levers and gears and other simple machines. They soon found that one worker with a lever and system of gears could raise the dam, which allowed the water to flow over a water wheel, which turned the grist mill and did the work of many serfs.

The change in medieval Europe was more gradual than the change we have experienced due to semiconductors, but the societal impact is comparable. Instead of a worker with a lever acting as a dam switch, we now have a battery in a circuit board switching currents off and on. Semiconductor switches, called transistors, are used for the on/off, zero/one logic that forms the basis of computers.

How can switches be used to generate all the many complicated functions of computers? First, by counting. Computers can count using just zeros and ones, which corresponds to "switch on" versus "switch off." Though we are used to counting in groups of ten, probably because we evolved with ten fingers, other systems are possible. Clocks count in groups of sixty (sixty seconds make a minute, sixty minutes make an hour) and eggs are packaged in groups of twelve—a dozen. Computers count in groups of two and use zero and one (switch off and switch on) for the digits. Computers also use AND, NOT, and OR, the same words that are used to link keywords in an Internet search (a system called *Boolean logic*) and if/then logic. For example, if the switch is on for the letter A, AND the switch is on for the printer, then the printer will print the letter A.

But if switches are all that is needed for computers, then why can't computers be built without transistors? The answer is that they could be, and they were. Switches are nothing new and even electronic switches are nothing new. In the past, vacuum tubes were used for the on/off switches (the zeros and ones) instead of transistors. Though they have other advantages and are used in other applications, the problem with using vacuum tubes in computers is that they are large and they create a lot of heat. The early computers required a lot of space and a lot of cooling. For any task of reasonable complexity, a great number of switches were required. With transistors, a very large number of switches can be put in a very small area without generating very much heat.

Once the basic principles of semiconductors were worked out, scientists directed their efforts toward devising different kinds of semiconductors. Nowadays there are semiconductors that sense heat or various colors of light. There are devices based on semiconductor logic that do everything from timing lightbulbs to piloting space craft. The construction of these various materials requires a confident command of the principles of chemical bonding and of the basic principles of chemical reactions, two of which—the conservation of mass and the law of definite proportions—are demonstrated next.

## Demonstration 7: Gemstone Chemistry

*"However," said she, "chemical laws exist." For, being a woman, she had need of belief.*
—Anatole France, *Penguin Island*, ca. 1900

Get two sandwich-sized self-sealing plastic bags and two gallon-sized self-sealing plastic bags and put on your safety glasses. Get a coat hanger and find a place to hook the hanger so it is suspended and can behave as a balance. A broom handle supported by two chairs might be the best arrangement for suspending the hanger. Into each large bag, place one level plastic teaspoonful (5 milliliters) of cream of tartar and one-half cup of hydrogen peroxide (60 milliliters). Mix these ingredients well by kneading the outsides of each bag with your fingers. Do not seal the bags yet.

Into each small sandwich-sized bag, pour a teaspoon (5 milliliters) of copper sulfate solution (prepared as described in the "Shopping List and Solutions"). Seal these bags completely, and place one into each of the two large bags. Now seal the two large bags and suspend them with clothes pins from the coat hanger, one on each side. Adjust their positions on the coat hanger by moving the clothes pins back and forth until the bags balance each other and the coat hanger is level. The homemade balance will work best if the entire coat-hanger assembly is hooked over something round so that the hook of the hanger can rock back and forth in an unrestricted manner. Once you have the bags balanced, secure the position of the clothes pin with adhesive tape. Check to see that the bags are still balanced.

Now comes the tricky part. In one of the big bags, open the little bag from the outside, without opening the big bag. Pour the copper solution from the little bag into the cream of tartar–hydrogen peroxide mixture in the big bag. The reaction begins slowly, but within ten seconds it should be fizzing away. The dramatic sky-blue copper tartarate eventually gives way to a lime-green mixture of the tartarate compound with orange copper oxide. The reaction generates enough heat that you can observe condensation on the inside of the big bag and feel warmth on the outside. The impressive colors produced by the reaction are reminiscent of gemstones for a very good reason: the colors of gemstones such as azurite and turquoise derive from copper salts. Various copper salts appear as dashes and streaks of color in polished stone.

Heat, bubbles, and color change—these are all indications that a reaction has occurred. The material now in the reaction bag is no longer the original material. However, if you were careful while opening the inside bag, the hanger assembly should have remained in balance. This balancing act is an example of *conservation of mass*: the idea that mass is neither created nor destroyed by chemical means. The mass of the reagents, the materials in the unreacted bag, remains the same as the mass of the products, the blue and green materials in the bag that were allowed to react. No matter how things change, some things just stay the same.

The second part of this demonstration has already been performed. Recall that the phenol red swimming pool pH indicator turned reddish orange in water and pink in baking soda solution. Recall that the vinegar-iron solution turned red when ammonia and hydrogen peroxide were added. These observations indicate that there is a difference between water, hydrogen peroxide, ammonia solution, and vinegar solution. What is the difference? Water has the formula $H_2O$, two hydrogen and one oxygen. Hydrogen peroxide has the formula $H_2O_2$, two hydrogen and two oxygen. The active principle in bases such as bicarbonate solution is the hydroxide ion, $OH^-$, one hydrogen, one oxygen. The active principle in acids such as vinegar is the hydronium ion, $H_3O^+$, one oxygen and three hydrogen.

There you have it: $H_2O$, $H_2O_2$, $OH^-$, $H_3O^+$. Not much difference typographically. But chemically, they're worlds apart, which exemplifies a principle called the *law of definite proportions.* It is not just the elements that determine the nature of a compound but the proportions, too. The ramifications of the conservation of mass and the law of definite proportions are the subjects we tackle next.

# CHAPTER 7

## Sticking to Principles

*The meeting of two personalities is like the contact of two chemical substances: if there is any reaction, both are transformed.*

—Carl Gustav Jung, ca. 1920

Some people are just naturally good cooks, adding a dash of this and a pinch of that. For the rest of us, there are recipes. We like explicit directions that spell out the required ingredients, the required amounts, the steps in producing the desired product, and number of servings that will be produced. While there may be some popular notion that chemists fit in the group that adds a pinch of this and a dash of that—and a good deal of creativity and art can enter into chemical research—the second group is where they spend most of their time. Balanced chemical equations, such as $2\ H_2 + O_2 \xrightarrow{spark} 2\ H_2O$, are chemical recipes.

We will begin our discussion of these recipes by examining two of the main principles on which they are based: the conservation of mass and the law of definite proportions. The law of definite proportions defines our ingredient list and the conservation of mass tells us how big a cake we'll bake.

The law of definite proportions tells us that our ingredients have a specific chemical formula; for example, $H_2O$ is the chemical formula for water. In essence, the chemical formula consists of a list of the elements that built the substance and the numbers of atoms of each that were required. There are subtleties, however, in the arrangement of elements, just as there are subtleties in the arrangement of elements in the periodic table. More information can be gleaned from the chemical formula than just the number and type of components.

For instance, the formula that we use for table salt, NaCl, always appears in the same order, sodium first and then chlorine. This order is based on a general agreement among chemists that the positive ion of an ionic compound should be named first. Because of the tendency of metal ions to form positive ions, metals are generally named first, followed by the nonmetal. It has also become standard practice to place the center element of a covalently bonded group at the head of the group. Thus ammonia—which has three hydrogens bonded via covalent bonds to a central nitrogen—has the formula $NH_3$. Carbon dioxide, whose formula is $CO_2$, has a central carbon with oxygens on either side. Looking at the formula, we know there are two oxygen nuclei involved because of the subscript two on the oxygen.

Moreover, the order of the elements in a chemical compound is important; thus, $CH_3CH_2OH$ is distinguished from $CH_3OCH_3$. Both substances have two carbon nuclei, six hydrogen nuclei, and one oxygen, but the first material is ethyl alcohol, sometimes called grain alcohol, and the other is dimethyl ether, which has decidedly different properties. Compounds that have the same number and type of elements but arranged in a different order are called *isomers*.

There are the three famous isomers: fulminic acid, cyanic acid, and isocyanic acid. Fulminic acid, HCNO, is used to make the explosives that share their name with the verb *to fulminate*, meaning "to verbally rage." Cyanic acid has the formula HOCN and is used to make poisonous cyanates. Isocyanic acid, HNCO, is less familiar because it calmly goes about its business as a starting material for making organic compounds and doesn't explode or poison you outright.

The law of definite proportions states simply that the formula for any one material is set and invariable. Any change to the chemical formula results in a new material with new properties (just as altering the ingredients in a batter can mean the difference between a layer cake and a pancake). When there is a formula-altering reaction, the recipe is given by a chemical equation. By convention, chemists place the beginning materials, the reactants, on the left, followed by an arrow, and the ending materials, the products, on the right. For instance,

$$C + O_2 \rightarrow CO_2$$

which means carbon plus oxygen will yield carbon dioxide.

There is a difference, however, between cookbook recipes and chemical reactions: not all chemical reactions go to completion. For many chemical reactions, a certain amount of the reactants remain in their unreacted state, like pudding that is imperfectly set. A pudding that is not properly set will have some portion that is pudding and some portion that is liquid. Most chemical reactions are like imperfect puddings: some portion of the reactants turn into products, but some other portion does not. The conservation of mass holds just as true for these not-perfect-pudding reactions, but for the sake of illustration, we will initially assume that all reactions go to completion; that is, we will assume all of the reactants turn into product, a pudding perfectly set. When a reaction goes to completion the amount of product is wholly predictable from the conservation of mass: if two pounds of reactants go into the recipe, two pounds of product come out.

The conservation of mass may seem like a statement of the obvious. If a chemical reaction occurs, such as the reaction in the above demonstration, we feel intuitively that the mass after the reaction should be the same as the mass before the reaction. The proof, however, is not always obvious. If a modern cook weighs all the ingredients for a pancake before cooking it and then weighs the pancake afterward, the cook is not surprised to find the mass is less. Bubbles can be seen during pancake cooking, an indication that a gas-phase product is escaping, and the pancake after cooking has a drier consistency, an indication water has been lost. But if this cook had been born at a time before the composition of air and water vapor were clearly understood, then the cook would know only that the pancake weighed less and would not necessarily conclude that mass had been conserved. One cook who was not fooled, however, was the chemist Lavoisier.

We met Lavoisier earlier, in conjunction with our discussion of redox reactions. We now add that Lavoisier was also instrumental in establishing the validity of the principle of the conservation of mass. He used the reaction

$$C + O_2 \rightarrow CO_2$$

but in true bourgeois style, he employed a diamond for his source of carbon and used sunlight focused through a huge lens to provide heat for the reaction. He carried out the experiment in a sealed vessel and showed by the inrush of air after the reaction that part of the air had been consumed. Before breaking open the vessel, however, he weighed it and showed that it weighed the same after the diamond-burning reaction as it weighed before: mass had been conserved.

Before Lavoisier's work could be translated into modern chemical reactions, several other important advances had to be made. Prominent among these was John Dalton's description of atoms. Though the idea had been kicking around since at least the time of the ancient Greek philosophers, Dalton took the important step of proposing that the atoms of different elements were different, and one of the characteristics that made them different was mass. Each element had atoms of different characteristic mass. This notion cleared the way for assigning a mass—an atomic mass—to each of the elements. In the above reaction between diamond and oxygen, the chemical formulas imply that a molecule of carbon dioxide forms when two oxygen atoms combine with one carbon atom. Dalton's atoms also assure us of the conservation of mass: there is one carbon and two oxygens on the left, and there is one carbon and two oxygens on the right—the same number of atoms, with the same mass, just arranged differently. If the number and type of atoms is the same on the left as on the right—the reaction is said to be *balanced*.

The reaction might be compared to the work of a florist. If a florist were to put a vase on the scale and a bunch of flowers beside the vase, we would not expect the scale's reading to change after the florist had picked up all the flowers and arranged them in the vase. Chemical reactions are like new floral arrangements—not new flowers, just new arrangements—so the mass is the same before and after the rearrangement of the chemical reaction.

Balanced chemical reactions convey essential information concerning how much reactant we will need to produce a given amount of product, or, conversely, how much product will be produced for a given amount of reactant. This type of information is important when one is dealing with reactions in beakers on benches. But when dealing with tank loads of chemicals in silo-sized reactors, this type of information is critical. Fortunately, chemical industries have chemists devoted to the study of the best ways of handling all sorts of tricky chemical situations from large quantities of heat, to large quantities of materials, to acid, to base, to gas-phase reactions: they are the intrepid chemical engineers.

## For Example: Engineering—Chemical, That Is

When people think of the chemical industry, they may think of the manufacturer of pesticides, pharmaceuticals, solvents, lubricants, or other relatively complex processes. Many people are surprised to learn that the vast bulk of the chemical industry revolves around the manufacture of four relatively simple chemicals: sulfuric acid, phosphoric acid, sodium hydroxide, and our old friend, sodium chloride.

Of these, sulfuric acid and its precursors are the largest volume chemicals produced in the world. Sulfuric acid has a syrupy appearance in its concentrated form (known as *oleum*) and was once known as *oil of vitriol* because of this appearance. Sulfuric acid is a highly corrosive acid that will attack plastic, wood, skin, mucous membrane, and most metals in its concentrated form, but it can be stored in glass. Very dilute sulfuric acid (a small amount of acid mixed with a large amount of water) is the effective ingredient in the aquarium pH-lowering solution suggested for purchase in the "Shopping List and Solutions." Sulfuric acid in a more concentrated form is the acid solution used in automobile lead-acid batteries. Nowadays automobile batteries are sealed in such a way that the average automobile user has no reason to check or fill the battery with fluid. A typical driver can enjoy an extensive motoring career without ever encountering sulfuric acid from a battery. But for those who have had contact with battery acid, the contact is memorable. The sensation to the skin is immediate and the effect on clothing is profound and irreversible.

The use of sulfuric acid in automobile batteries, however, does not account for its production in such vast quantities. The largest single use for sulfuric acid is fertilizers, especially a sulfuric-acid/phosphate-containing-rock mixture called superphosphate. Because phosphorus is a major mineral nutrient needed by plants and because phosphorus is a major component of bones, bone meal was once a commonly used fertilizer. Today, however, phosphate rock, treated with sulfuric acid to become more soluble, can be used, too.

Sodium hydroxide or lye, NaOH, is used in the production of soap, textiles, petroleum products, dye, detergent, and paper. Sodium hydroxide is produced by running an electric current through a mixture of water and NaCl (table salt). The balanced equation is straightforward:

$$2\ H_2O + 2\ NaCl \rightarrow 2\ NaOH + Cl_2 + H_2$$

The number in front of the formula for water and also in front of the formula for table salt indicates the number of units of each that enter into the reaction. Two water molecules react with two units of sodium chloride to form one unit of sodium hydroxide, one molecule of chlorine gas, $Cl_2$, and one molecule of hydrogen gas, $H_2$. In other words, if we were to write each reactant individually, the reaction would be

$$H_2O + H_2O + NaCl + NaCl \rightarrow NaOH + NaOH + Cl_2 + H_2$$

The tally of hydrogen before the reaction is two from each water, or four hydrogens. The total of hydrogen after the reaction is two from $H_2$ plus one from each NaOH, or four hydrogens. Thus we have the same amount of hydrogen before and after reaction but found in different compounds. The hydrogen is balanced. The tally of oxygen before the reaction is one from each water, or two oxygens; the total oxygens after the reaction is one from each NaOH, or two oxygens. The oxygen is balanced. The tally of sodium, Na, before the reaction is one from each NaCl; the total for sodium after the reaction is one from each NaOH. The sodium is balanced. The total of chlorine is one from each NaCl before the reaction and two from $Cl_2$ after the reaction.

For the chemical engineer, however, the problem with this particular method for producing NaOH is not balancing the reaction but dealing with the by-products. The by-products of this reaction, $Cl_2$ and $H_2$, are both gases under normal conditions, and gases occupy more space, pound per pound, than their solid counterpart. The production of gaseous by-products was the basis of the bottle rocket demonstration. The baking soda reacted with the vinegar, creating a carbon dioxide gas that expanded to occupy a greater volume than the reactants and forced the cork to pop. For every ton of sodium hydroxide produced, there is about one ton of gas produced. For every cubic foot of sodium hydroxide, NaOH, produced, over one thousand cubic feet of gases are generated. This volume difference would be like a seventeen-inch television set swelling up to fill two rooms of a house. That's a lot of gas and it could cause a very big pop. Thus containment, ventilation, and emissions pose major challenges for those in the chemical-producing industries.

But just as an underestimation of such problems could cause a chemical engineer's career to falter, so would an overestimation that caused unnecessary expenditure. Therefore, the chemical engineer must also be aware that an estimate, such as the one given above, is based on the assumption that each molecule of gas occupies a volume as though it were all alone in the world. In actuality, there are attractions and repulsions between the molecules. The attractions between molecules are why gases can be made to condense to liquids. The repulsions between molecules are what make materials more and more difficult to compact as the volume becomes smaller and smaller. These attractions and repulsions, collectively called the *intermolecular forces*, have to be taken into account by the chemical engineer. These intermolecular forces are also the basis for many other intriguing observed properties of materials, which is what we explore next.

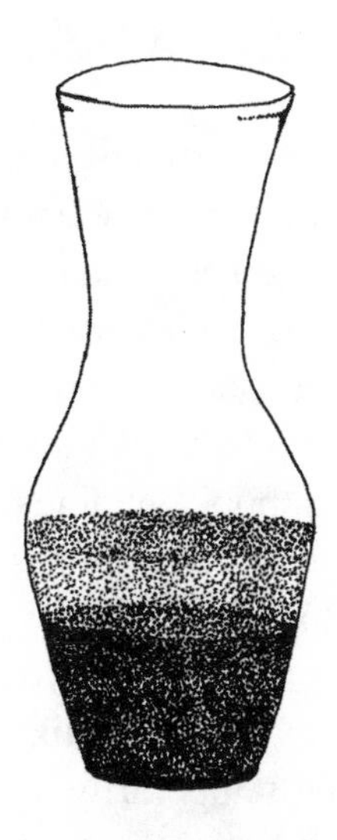

## Demonstration 8: Layer upon Layer

*Now if she meet a man—suppose our hero,*
*With whom her chemistry shall war yet mix,*
*As if she were her Borgia to his Nero,*
*'Twill look like one of Satan's little tricks!*
—Edgar Lee Masters, *Spoon River Anthology,*
ca. 1916

After putting on safety glasses, take a small clear glass and pour in glycerin (suggested for purchase in the "Shopping List and Solutions") until the glass is about an inch (2.5 centimeters) full. Take some water dyed with food coloring and gently pour about the same height onto the glycerin. You should observe two distinct layers in the glass.

Add a layer of canola oil to this and observe the three distinct layers. Stir the mixture. The colored water will mix with the glycerin, but the canola oil will still settle into a separate layer once agitation has ceased. It may take several minutes for the layers to separate.

Add a layer of liquid soap to the solution and watch it settle between the glycerin-water layer and the canola-oil layer. Now, when the combined layers are mixed (gently so as not to cause bubbles), the soap will cause the canola oil to blend with the glycerin. The result is a uniform mixture. The mixture may eventually reseparate into different layers, but it takes longer than when the detergent wasn't present.

The solutions initially separated because of a difference in *intermolecular attraction*, a type of *intermolecular force*. These attractions occur *between* molecules, as distinguished from *intra*molecular forces—chemical bonds—which bind the nuclei *within* a molecule.

The difference is one of book bindings versus sticky notes. Book bindings are like chemical bonds. Book bindings are meant to be permanent. The glue that is used to hold the pages together is not easily separated. Once the book binding is separated, you have a stack of pages; you no longer have a book. Sticky notes are like intermolecular attractions. Sticky notes are meant to be temporarily stuck to a surface and easily freed and moved. If you stick a sticky note to a book and then remove it, you still have an intact book and sticky note. Sticky notes stick, by the way, because of intermolecular attractions.

We have already encountered the effects of intermolecular forces in our discussion of precipitates and solubility. Here the intermolecular attractions between water molecules are instrumental in the ion-cage formation that allows some salts to go into aqueous solution. The glycerin molecule shares some similarities with water, but the individual glycerin molecules are still strongly attracted to each other and admit water to their ranks only when there is sufficient provocation. In this demonstration, the provocation occurs in the form of stirring, but no amount of stirring will force the canola oil into the glycerin solution until soap is added.

Soap is the grand diplomat of intermolecular forces. Soap gets along with and experiences attractions to both water and oil, which is the basis for its efficacy as a cleaning agent. Soap combines with grease, oil, and dirt, and then with water, which allows the dirt to be washed away with the water.

But soaplike molecules, molecules that have an attraction to both waterlike substances and oil-like substances, have a much more important role than cleaning skin: they form the walls of the very cells of the skin. Intermolecular forces may at first seem relatively unimportant when compared to chemical bonds, but while chemical bonds determine the structure of a single molecule, it is intermolecular forces that will determine how this molecule will behave around other molecules. And the behavior of molecules, as observed in the properties of molecular substances and mixtures, is what life is all about.

# CHAPTER 8

## Slipping and Sliding, Intermolecularly

*(H)e found himself at first gazing at the portrait with a feeling of almost scientific interest. That such a change should have taken place was incredible to him. And yet it was a fact. Was there some subtle affinity between the chemical atoms, that shaped themselves into form and colour on the canvas, and the soul that was within him?*

—Oscar Wilde, *The Picture of Dorian Gray*, ca. 1890

Intermolecular forces play an important role in chemistry. Intermolecular forces are the attractions and repulsions felt *between* atoms and molecules and differ from chemical bonds. Breaking a chemical bond creates a new material that reacts differently with other chemicals. Breaking an intermolecular attraction does not change the chemical formula or chemical reactivity. But intermolecular forces can determine how quickly a reaction will occur—or if it can ever occur. And intermolecular forces determine whether a material will be a liquid, a solid, or a gas at a given temperature—an important consideration for those of us who like to drink, eat, and breathe.

In the late 1800s and early 1900s, scientists were still struggling to understand intermolecular forces, so it is doubtful that Oscar Wilde had a clear picture of intermolecular forces in mind when he wrote of the "subtle affinity" between chemical atoms in *The Picture of Dorian Gray*. Nonetheless, his description of "subtle affinity" is quite apt. Intermolecular forces are complex, consisting of attractions as well as repulsions. Intermolecular *attractions* are those between water molecules that allow water to condense once it has been sufficiently cooled—and intermolecular *repulsions* are what make water feel like a solid mass when it is forcefully encountered. (Have you ever been knocked over by a wave?) If it were not for intermolecular attractions, our bodies would vaporize into gases, and if it were not for intermolecular repulsions, we would collapse into unimpressive puddles.

We begin with intermolecular repulsions. In the "Water Witch" demonstration presented earlier, we saw a thin stream of water being attracted to a plastic spoon with an acquired static charge. Similarily, if a piece of adhesive tape is yanked rapidly from a roll, it too will attract a thin stream of water. However, if two pieces are yanked from the roll and dangled from the fingers, these pieces will attempt to escape from each other. They acquired the same charge when they were removed from the roll of tape and are repelled because like charges repel. When two atoms or molecules are attracted to each other, they will come only so close and no closer. Electrons have intimacy issues. When two distinct atoms or molecules come too close together, the repulsion between the electron clouds will keep them apart.

Intermolecular attractions are also electrical in nature, but the situation here can be more complicated because intermolecular attractions can depend more on the shape of the molecule, a shape that is constantly undergoing slight changes. Molecules are in constant motion—vibrating, rotating, and flying off in all directions. The problem is one of catching a boomerang as opposed to a baseball. Moreover, molecules do not move at the same speed but in a range of speeds. They are not all rotating or vibrating at the same frequency but with a range of frequencies.

These variables make a straightforward calculation of the magnitude of the intermolecular attractions in any one situation virtually impossible. But despite their individuality, molecules do stick together. And understanding this stickiness is critical to the understanding of the behavior of materials. As a start to unraveling this sticky business, we identify three ways molecules can be attracted to each other: ion-to-dipole attraction, dipole-to-dipole attraction, and a newcomer on our horizon—dispersion forces.

Positive-ion-to-negative-ion attractions are straightforward and strong interactions: so strong, in fact, that they tend to dominate when they are present. As we pointed out in our discussion of precipitation, when opposite ions find each other, they can form an ionic bond and come out of solution. Ionic attractions do not, however, always prevail. Water molecules can keep ions apart by forming cages around the ions. These water cages are the result of dipole-to-dipole and dipole-to-ion attractions.

To better understand dipoles, recall that earlier we explained that electrons will tend to associate themselves with the more electronegative entity in a bonding situation and that this can cause an unequal distribution of electrons. One end of a bond is more negative or more positive than the other. When this is the situation, we say that the bond is *polar*. When polar bonds are directly opposite from each other, they can offset, or cancel, each other, and though the bonds may be polar, there is no polarity for the molecule as a whole. This is the case with carbon dioxide, $CO_2$. In carbon dioxide, the electron density is shifted toward the more electronegative oxygens, but carbon is seated in the center of two oxygens, as OCO, so the tug-of-war is a stalemate.

Water, however, is a different story. As noted earlier, the water molecule, $H_2O$, is shaped in a **V** with the oxygen in the middle and the hydrogens at the tips of the **V**. As a result, water is a polar molecule. Oxygen is more electronegative than hydrogen, so the electrons gravitate toward the oxygen end and away from the hydrogen end. Water is said to have a *net dipole moment*, meaning that when all the forces have balanced out, one end is more negative than the other.

As we've seen, polar molecules interact with each other somewhat as bar magnets do, and the interaction is termed a *dipole* interaction. In a dipole interaction, the positive end of one polar molecule lines up with the negative end of another, as the north pole of one bar magnet will line up with the south pole of another—but these magnets are rotating, vibrating, and flying in all directions.

Earlier we were interested in how this interaction helped to build cages around ions to keep them in solution, but the interaction is important in the absence of ions, too. Dipole interactions serve to hold water together in the condensed state. Water is especially adept at dissolving and condensing because water molecules are small and able to experience yet another type of intermolecular attraction. A strong dipole/dipole attraction called a *hydrogen bond.*

A hydrogen bond is not a chemical bond in the sense that breaking a chemical bond causes a change in the molecular formula. So the term hydrogen *bond* is a bit of a misnomer, but a hydrogen bond is a particularly strong attraction. Hydrogen bonds occur when a hydrogen is *chemically* bonded to a very electronegative element, such as oxygen, fluorine, or nitrogen, and attracted—via intermolecular attraction—to an electronegative element on another molecule. The situation is shown in figure 1.8.1. Hydrogen bonding is also an especially important factor in the behavior of water because water has four opportunities for hydrogen bonds: it has two hydrogens chemically bonded to the oxygen, and the oxygen can attract two outside hydrogens.

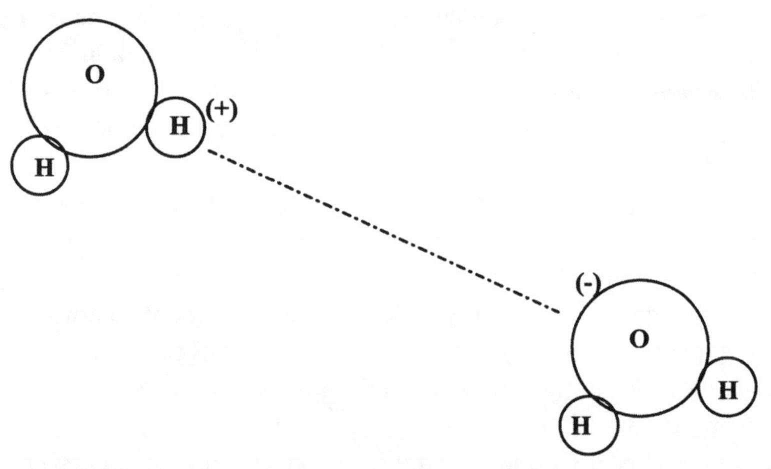


Figure 1.8.1. The attraction between the slightly positive hydrogen end of one water molecule and the slightly negative oxygen end of another is called hydrogen bonding. Hydrogen bonding is an intermolecular attraction, not a chemical bond. To experience hydrogen bonding, the hydrogen must be chemically bonded to a highly electronegative element such as oxygen, nitrogen, or fluorine.

Because it can form four hydrogen bonds, water is very talented at forming cages and dissolving ions. Because of this special ability of water to dissolve, water is the vehicle for many key chemical reactions, including those responsible for life. Many liquids we deal with on a daily basis are either water or water solutions.

Another common liquid is liquid oil, either in the form of cooking oil or gasoline or other petroleum product. Not only are these materials definitely not water, but they do not mix very well with water. Oils and petroleum products are composed mainly of hydrogen on long chains of carbon, with a smaller amount of oxygen and their intermolecular attractions are different from those of water.

At first it might be thought that the bonds in carbon–hydrogen chains would be polar, too. Carbon and hydrogen aren't very close to each other on the periodic table and therefore should have a significant difference in electronegativity. But as you will recall, hydrogen does not follow the general trend—it is too small to carry much of a negative charge—so hydrogen is actually very close to carbon in electronegativity. In addition, the hydrogens can be distributed evenly around carbon in carbon–hydrogen chains, so the slight polarity of the carbon-hydrogen bond tends to be balanced out. But these chains obviously do form liquids, and even solids called waxes, so there must be some form of intermolecular attraction.

Condensation of carbon–hydrogen chains can be better understood when it is realized that it is usually the larger chains that condense more readily. Methane, composed of one carbon and four hydrogens, is a gas (swamp gas to be exact) at normal temperatures and pressures. Octane, composed of eight carbons in a chain and eighteen hydrogens, is a liquid under normal conditions. The longer chains are more apt to be liquids at room temperature and pressure because the long chains become intertwined. Tangled up, at close range, even the weak attractions between the slightly polar bonds can contribute to the stickiness, too.

On the other hand, nitrogen gas, $N_2$, can have no difference in electronegativity because two identical atoms are joined, so the bond cannot be inherently polar. And there is no opportunity for tangling because there are no chains of nitrogen, just two-atom pairs. Yet liquid nitrogen exists, and it is common enough to be used for wart removal. What holds nitrogen molecules together in liquid nitrogen? An attraction called a *dis-*

*persion force* or an *induced dipole attraction* or a *London force*, named after the scientist Fritz London, who described it.

Whichever name it is given, the origin of this attraction is the mushy electron cloud that surrounds the nitrogen molecule. Because the electrons can be considered mobile in the electron cloud, they can be pictured as congregating momentarily at one end of the molecule or the other. This momentary uneven distribution of electrons is termed a *temporary dipole*, but it acts in the same manner as a permanent dipole. It is attracted to other dipoles, temporary or otherwise. The redistribution of electrons may be spontaneous, or if there is an ion or a molecule with a permanent dipole in the vicinity, this species might induce a momentary dipole, too. This situation is shown in figure 1.8.2.

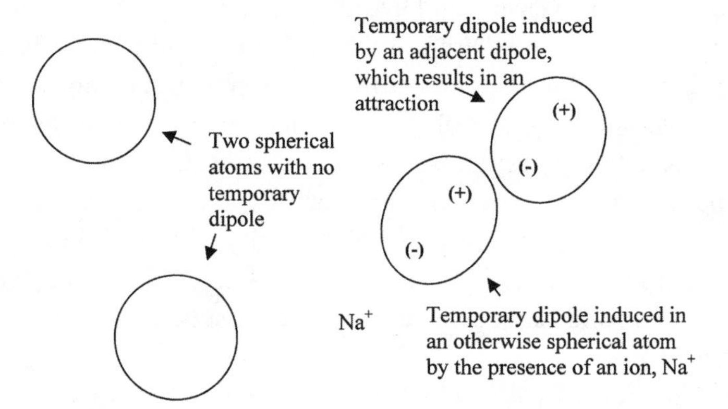


Figure 1.8.2. The sodium ion is inducing a temporary dipole in an atom with a normally spherical electron cloud. The induced dipole is causing another atom to have a momentary dipole. These temporary dipole attractions are sufficient to cause even noble gases to condense at low temperatures.

Just as the temporarily charged spoon in the "Water Witch" demonstration caused the water to bend toward it, an induced dipole in one species can induce a dipole in another. The dispersion forces aren't as effective over distance as dipole attractions, but they are sufficient to account for the existence of liquid helium as well as liquid nitrogen and liquid oxygen.

Helium, however, does not condense nearly as readily as water because helium has only one intermolecular attraction at work. Water has an accumulation of intermolecular attractions. In saltwater there can be all permutations of interactions: dipole/dipole, dispersion/dispersion,

ion/ion, dipole/dispersion, dipole/ion, and ion/dispersion. So the situation, as we noted, is quite complicated. Added to this confusion is the fact that the attractions fall off with distance, which is always changing because of the many modes of motion of the molecules, and the situation becomes quite desperate.

Fortunately, we do not have to dissect each interaction to understand the net effect—one that manifests itself in several well-known phenomena such as friction, cohesion, adhesion, capillary action, solubility, and the action of soap.

Frictional forces are perhaps the most complex and bedeviling forces we have to deal with on a day-to-day basis. Frictional forces arise from two sources—physical barriers to motion and intermolecular forces—and for this reason can never be completely defeated. The physical barriers to motion are the microscopic ridges and rough spots that are always found if you look closely enough. When an attempt is made to drag one material past another, the rough spots of one side can catch in the rough spots of the other side and cause friction, or resistance to motion. Removing the rough spots helps, but only to a point. When materials become very smooth, there is greater opportunity for intermolecular attractions.

Grease sometimes helps reduce friction because grease forms a film that keeps the moving materials apart. The friction within the grease, as it is pulled one way or the other, eventually causes the grease to decompose, which is why bearing grease in a machine must be periodically replaced, as well as oil and lubricants in automobiles.

*Cohesion* is the tendency of like molecules to stick to one another. For instance, oil likes to stay with oil and water with water, which is why they separate when allowed to stand. Surface tension, manifest in the curved rise of water over the top of a glass or spoon, is the result of cohesion. Water, because of its special propensity to form hydrogen bonds, displays a good deal of cohesion and therefore surface tension. The surface tension of ponds is strong enough to support a world of water-walking creatures.

Adhesion, as in adhesive tape, is the tendency of different materials to stick together. Adhesion is a major factor responsible for surface waves in bodies of water. Although the physics is a complex trade-off involving many forces, waves begin by water sticking to the wind blowing past. The idea of pouring oil on troubled waters, often used as a metaphor for a calming effect, has a physical basis and was actually used with limited success by whaling vessels in the past. Oil will disrupt the adhesion between water and wind and can calm some waves caused by wind. Of course, large waves that have already formed and are being carried forward by their own momentum will not be affected. So don't take the time to look for the bottle of salad oil when the tsunami is coming—just run.

There is a good deal of adhesion between molecules with similar structure, such as water and rubbing alcohol, and water and grain alcohol, the structures of which are shown in figure 1.8.3.

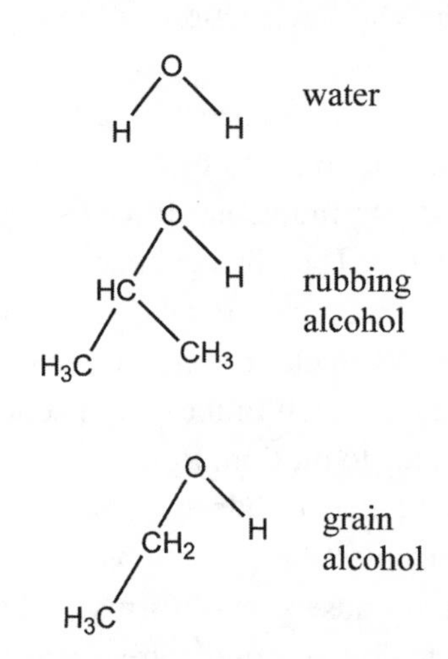


Figure 1.8.3. A comparison of the structure of rubbing alcohol, grain alcohol, and water.

The mutual solubility of water and alcohol is the reason a highball doesn't normally separate into layers. Some syrups, such as grenadine, may take longer to mix because of the intermolecular attractions within the syrup itself. But if one waits long enough, the syrups in fancy layered drinks eventually distribute throughout the drink. But the point is usually drinking, not waiting, so the effect may go unobserved.

Adhesion is the basis for *capillary action*, or the tendency of liquids to travel against the pull of gravity. The molecules of the material traveling upward have a greater attraction for the surface on which they move up than the bulk of the liquid being pulled down by gravity. Paper towels work on the principle of capillary action, which is why paper towels work better on some liquids than others. Try soaking up water versus glycerol or cooking oil by making a puddle of each and dipping in the corner of a paper towel. The water travels up the paper towel much more quickly than the oil does. Water is more attracted to the molecules of the paper towel. Oil molecules are more attracted to each other.

Adhesion is also the basis for the adage *like dissolves like.* "Like dissolves like" means oil-like liquids dissolve oil-like liquids and waterlike liquids dissolve waterlike liquids. "Like dissolving like" explains why adding an antistatic dryer sheet to the water when soaking a pot with baked-on grease will help to remove the greasy material. These sheets add a very thin coating of oil to the clothing being dried, which prevents friction from stripping off electrons and causing a static charge. This same nonpolar oil can combine with the nonpolar baked-on material and help lift it free of the pot. In the same manner, cooking oil can help to remove bandages from skin without exfoliating the skin. The oil combines with the oil-like materials in the glue. "Like dissolving like" also explains the action of soap.

Soaps are made from long carbon–hydrogen chains with an oxygen-containing group attached at one end. The end with the oxygen-containing group is soluble in water because that end is polar and able to form hydrogen bonds. Other carbon chains, like grease, for instance, are soluble in the carbon-chain end of the soap and are carried into water solution by the water-soluble end. This same type of molecule, one that is water soluble on one end and grease soluble on the other, is vital to the structure of cells and other life-giving chemistry. For instance, the body needs many nonpolar substances to be transported through water-based blood, and molecules that straddle both worlds—polar and nonpolar—aid in this transportation. Still unimpressed with the importance of intermolecular forces? Then consider this: without intermolecular forces, there could be no flush toilets.

## For Example: The Physics of the Flush Toilet

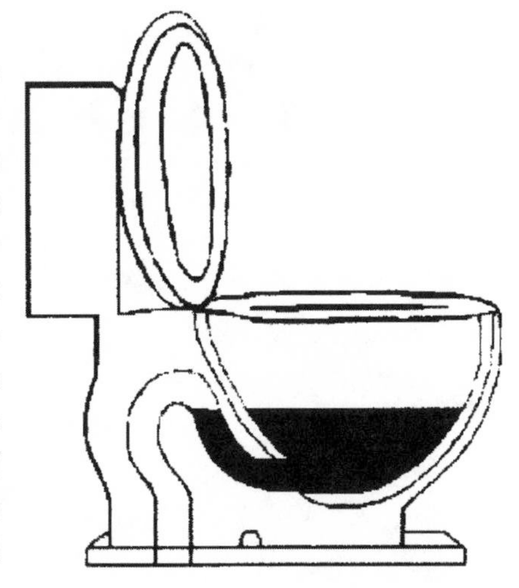

Flush toilets are easily one of the most significant advances of civilization. Not only do they remove an unsightly and potentially dangerous breeding ground for bacteria, they do so in such a civilized manner: waste is removed quickly and efficiently, and the odors that would be associated with the stored waste are cleverly kept from us by a wall of water. The flush toilet pushes waste up and over an inverted-U-shaped hill and blocks the entrance with water so that sewer gases will not find their way back into the room. None of this would be possible if it were not for intermolecular forces.

The flushing action of the toilet commences when the bowl is filled from a reservoir until the water level tops the inverted-U-shaped trap and starts to spill over the other side. Once water starts to flow over the side of the trap, it pulls the rest of the water with it, as a chain would pull the links that are attached to it down to a lower level. With water, intermolecular attractions form the chain that pulls the water molecules one after another. This siphoning action has to initiate quickly so that water will surge over the trap and make a moving seal.

The difference in pressure between the filled bowl and the empty pipe shoves the water over the top, but it is the attractions between the water molecules that keep the flow moving along and continues through siphoning action to drain the bowl. That intermolecular attractions are involved in the flush toilet can be confirmed with a thought experiment: can you imagine a toilet with sand as the working fluid? This medium might be a kitty's dream, but it won't work with a flush! Siphon action depends on one molecule dragging along another once it is over the hump.

Water also works well for waste disposal because so many materials dissolve in water. But as we all know, there is a big difference between toilet water—as in water in the toilet—and toilet water, as in eau de toilet. The properties of the solution depend on the solvent, the solute, and the amount of solute it contains. Discussion of these important parameters is where we are headed next.

# Demonstration 9: Concentrating on Color, Salty Dog

*Like a chemist, Napoleon considered all Europe to be material for his experiments. But, in due course, this material reacted against him.*

—Frédéric Bastiat, *The Law*, ca. 1850

Put on your safety glasses. Pour one-half cup of water (120 milliliters) into each of two clear plastic cups. Add a plastic teaspoon (5 milliliters) of copper sulfate solution (prepared as outlined in the "Shopping List and Solutions") to one cup and two plastic teaspoons (10 milliliters) of copper sulfate solution to the other cup. Stir the two glasses and compare the intensity of the color by looking down from the top of the two glasses while they are on a white piece of paper. The color of the glass with one teaspoon of copper sulfate solution (the less-concentrated solution) should appear several shades lighter in color than the glass with two added teaspoons of solution. Color is a property influenced by concentration.

Solubility is influenced by the presence of other materials, as discussed in conjunction with precipitation. A few pinches of baking soda go into solution readily, but not much more than that. Now try the following.

Make a saturated baking soda solution by adding four tablespoons (60 milliliters) of baking soda to two cups (480 milliliters) of water. Stir the solution and allow it to settle for at least four minutes. Carefully decant the liquid over the undissolved baking soda into a clean glass, making certain no solid is transferred with the liquid. Take an identical clean glass and fill it with water to the same height as the baking soda–water glass. When you have two glasses with an equal amount of liquid, add a level quarter teaspoon (1 milliliter) of table salt to each glass and stir. While stirring, keep adding the same amount of table salt, in small increments (eighth teaspoon; half milliliter), to each of the glasses until one of the solutions turns cloudy. The cloudy solution will be the one with the baking soda. It turns cloudy because the baking soda is crowded out of solution by table salt. The other glass, the one with just water and added salt, should be completely clear. This technique, called *salting out*, is used industrially to move materials in and out of solution. Salting out was also used historically in the preparation of solid soap. The soap is forced out of solution by adding large amounts of salt. This effect can be demonstrated by pouring an eighth- to quarter-inch layer of liquid soap into a glass and sprinkling table salt on top. In a few minutes a nice, gooey, gelatinous solid will precipitate out.

# CHAPTER 9

## Concentration—On Being Alone Together

*There was no trace then of the horror which I had myself felt . . . but his face showed rather the quiet and interested composure of the chemist who sees the crystals falling into position from his oversaturated solution.*

—Arthur Conan Doyle, *Valley of Fear,* 1915

Chemistry is change—ore to metal, batter to bread, acid to salt—and a good deal of this change, this chemistry, occurs in solutions: ponds, cells, oceans, sauces, melts, and mixtures. Solutions are uniform mixtures, such as the uniform mixture of salt and water we call saltwater. Solutions are particularly conducive to chemical reactions because they provide the opportunity for molecules to come together and the mobility for the reaction to take place. The tablets that are dropped into water to make a fizzing tonic are able to make a fizz because they contain citric acid and baking soda. But even though the citric acid and the baking soda are present in the tablet before it is dropped into water, the dry tablet does not fizz. It will not fizz until the ingredients are in a solution. In solution, the acid and base have mobility and can react. Solutions provide what's needed for chemical reaction: mobility and opportunity.

Successful commercial and intellectual enterprises have long understood the need for mobility and opportunity for productive interaction among specialists. At conferences, people of like mind are brought together in high concentration. But as any successful conference organizer knows, the real chemistry doesn't happen when everyone is stationed in their seats listening to presentations—it happens at the wine-and-cheese tasting when people have the mobility to seek each other out. A solution is the chemical equivalent of a conference: a concentration of reactants with the mobility that they need to combine. A chemist, not unlike a conference organizer, understands the importance of controlled concentrations and orchestrates them when designing reactions.

Though water solutions, or *aqueous solutions* as they are known to the chemist, are probably the most familiar solutions, there are many other kinds of solutions. There are solid solutions such as metal alloys, although not all alloys form truly homogenous blends. Bronze is an alloy of copper and tin, brass is an alloy of copper and zinc, and steel is an alloy of iron and carbon.

There are also gas-phase solutions. The air we breathe is a solution of several gases. Nitrogen is present in the greatest proportion, about 80 percent, but oxygen is also mixed in, as well as argon, carbon dioxide, and usually water. The amount of water in air can vary from none to 100 percent humidity, while the amounts of nitrogen, oxygen, and argon stay fairly fixed. The presence of argon in the solution of gases we breathe may be a little surprising at first. But recall from our discussion of the periodic table that the lighter elements were those first produced in the cosmic porridge from which our planet emerged. Reactive elements were trapped in compounds; unreactive elements floated away. Most of the helium and neon were light enough to escape Earth's atmosphere, which left argon to represent the noble gases. Other gas solutions include carbonated soda, a solution of a gas (carbon dioxide) in water, and pond water, a solution of oxygen in water—which is how fish breathe.

To carry out metabolic functions, the body depends on the solubility of gases in blood, which, in turn, depends on the pressure *of that gas* and not any other. So the solubility of oxygen in blood depends on the pressure of oxygen in the inhaled air, and the solubility of oxygen in the cells depends on the pressure of oxygen in the cell. As metabolism takes place in the cell, oxygen is used up and carbon dioxide is produced. When oxygen rich blood reaches the cell, oxygen passes from the blood—where the concentration is relatively high—to the cell, where it is relatively low. Carbon dioxide passes from the cell, where its concentration is high, to the blood, where its concentration is low.

Because nitrogen is also always present in the air we breathe, there is always some amount of dissolved nitrogen in our blood, too. Divers who descend deep into the ocean need pressurized air tanks to deliver their breathing air because the pressure of the ocean is pushing in on their lungs. The pressure inside the tanks must be greater than the external pressure if it is to enter the lungs. If normal atmospheric air is used to fill the tanks, the blood being restocked at the lungs will be exposed to nitrogen at higher pressures, and there will be more nitrogen dissolving in the blood. The extra nitrogen in the blood is not necessarily a problem, unless the diver comes up too quickly. If the diver does ascend too rapidly, the extra nitrogen that was dissolved in the blood starts to come out of solution and can form bubbles before the body has a chance to exhale it. This condition can cause such severe pain that sufferers have been known to bend over in pain, hence the name: the bends.

A preventative measure often employed is to substitute helium for nitrogen in the breathing air mixture. Helium is not as soluble as nitrogen in blood and therefore does not cause the same problem. On a completely different note, the strange, Donald Duck sound helium imparts to the voice is the result of the less dense helium, as opposed to air, passing through the larynx. There is no intoxication caused by helium, as sometimes thought, but inhaling pure helium can be dangerous. Someone inhaling pure helium is not inhaling oxygen and can pass out from lack of oxygen.

Solutions consist of two parts: the solvent and the solute (or solutes). The solvent is the component present in the majority, and the solute is the component present in the minority. In saltwater, the quintessential solution, water is the solvent and salt is the solute. The importance of concentration as a property of solutions can be understood by considering saltwater. Concentrated saltwater can be used to rinse the mouth, but only dilute saltwater can be used to rinse the eyes. You would not want to confuse the two. Concentration matters.

There are some common measures of concentration that have found their way into everyday use. The measure of the concentration of ethanol in solution in alcoholic beverages is called *proof*. Proof is two times the percentage of ethanol in solution, volume for volume. For example, a 76-proof whiskey is 38 percent alcohol. Legend has it that the term came from the fact that a mixture of about 50 percent ethanol (100 proof) and water will burn, proving it is good alcohol. Alcohol-water solutions have the interesting property that their volumes are not additive in all ratios. For instance, equal quantities of rubbing alcohol and water mixed together does not form a solution that is double the original quantity—a result of those attractive intermolecular forces we discussed in the last chapter. Alcohol has an OH group as part of its molecule, and this OH group is attracted to $H_2O$. A mixed group of single men and single women might be expected to take up less space than the separated groups for the same reason: attractive interactions.

Another common measure of concentration is humidity. Humidity is a measure of the amount of water in air, and relative humidity is the amount of water currently in the air compared to the maximum amount of water that the air could hold at that same temperature and pressure. When it comes to solutions of solids in water, such as saltwater or sugar water, the concentrations can be given in grams per milliliter, as they are in many common medicines. When it comes to chemical reactions, though, it is more meaningful to talk about molecules per liter than grams per liter. To understand why, consider the following two statements. Which is more useful for calculating staff requirements if the teacher/student ratio needs to be maintained at twenty to one?

There are 11,756 kilograms of students in the new freshman class.

or

There are 190 students in the new freshman class.

Chemists usually characterize the concentration of a solution by a measure of the number of molecules per volume rather than the number of grams per volume because reactions happen when one molecule reacts with one molecule, not when one gram reacts with one gram. The difficulty is that molecules are very small, and a good many of them are required to make up a measurable amount. For instance, it requires about a trillion trillion water molecules in the liquid state to fill a 240-milliliter (one-cup) container. Therefore, it is infinitely easier to speak of collections of molecules rather than individual molecules. The name that chemists agreed on for a collection of molecules is a *mole*, and a mole is like a dozen—only different.

A dozen indicates a number of things, such as, perhaps, twelve eggs. A gross indicates a number of things, 144 things. A mole also indicates a number of things: a mole is 6.023 hundred billion trillion things, an almost unimaginably huge number.

For instance, if molecules were as big as marbles, then a mole would cover the land area of the United States to a depth of four meters (about the height of the first story of a house). Cells in the human body are too small to be seen without a microscope, and so small that about one hundred trillion cells are crammed into one human body. But to assemble a mole of cells, you would need six billion people, the population of the entire Earth in 1995. A pretty big number.

It is necessary to use this large a measure when it comes to molecules because molecules are so small. A mole of eggs would be 6.023 hundred billion trillion eggs. That would make an overwhelming omelet, but it's really an underwhelming quantity of molecules. A mole of sucrose molecules is less than two cups (470 milliliters) of sugar: not quite enough for one cake. A mole of solid is about a handful, and a mole of gas is about six gallons (twenty-two liters) under normal conditions of temperature and pressure (a little more than a bucket and a little less than a bushel). In the appendix, the mass given for each element can be interpreted as atomic mass units per atom (amu per atom) or grams per mole. This is not just a fortunate coincidence. The magnitude of an amu was chosen so that amu per particle would correlate with grams per mole.

Earlier we mentioned that a common measure of acidity is pH, which is a measure of the concentration of hydronium ions ($H_3O^+$, the acidic agent) in solution. Now we can be more specific: a solution with a tenth of a mole of hydronium ions per liter has a pH of 1. A solution with a hundredth of a mole of hydronium ions per liter has a pH of 2. Every time the concentration of hydronium ions decreases by a multiple of ten, the pH goes up by one unit.

Another measure related to concentration is density. Density is a measure of the total mass of material per unit volume. For solutions in which chemical reactions are ongoing, measuring the density can be a way of measuring the progress of the reaction. Mass has to be conserved in the course of the reaction, but the volume that the mass occupies can change considerably due to rearranging molecules and rearranging intermolecular forces. Such variations in density are evident in bodies of water, the human body, and another interesting body: the body of wine.

## For Example: What Body, What Aroma, What Chemistry

The making of wine involves chemistry on many levels. For instance, winemakers and wine tasters rely on concentration properties to make and describe their . . . how shall we say . . . delicate, yet somewhat bold and enticing, product.

Professional wine tasters use a number of interesting terms such as *corky*, *buttery*, *chewy*, *fresh*, *fruity*, *honeyed*, *musty*, *round*, *complex*, *acid*, *thin*, *youthful*, *nutty*, *full*, or *zestful* to describe the attributes of wines, which include aroma, bouquet, nose, astringency, finish, texture, and body. It is this last attribute, body, that connects with the subject of this chapter: concentration. Body is the perceived density of the wine in the mouth and depends in good part on the amount of alcohol in the wine. Density, related to concentration, plays a vital part in wine making, too.

The *must* is the beginning mixture, the grape juice and crushed fruit ready to be fermented. Fermentation is the process wherein certain yeasts consume sugar and produce ethanol (alcohol) and carbon dioxide. Too little sugar leads to too little alcohol, and too much sugar changes the nature of the wine. The measured density of the *must* can be used to estimate the sugar content. From the measured sugar content, the alcohol content of the finished product can be predicted. On this basis, it can be decided if sugar needs to be added to the mixture. Because sugar is a reactant and used up, the density is also measured during the fermentation process to track how well fermentation is proceeding.

So, as we have seen, measurement and control of concentration is critical in chemistry as well as wine and good conferences. But the liquid state is not the only state in which chemistry is found. The gas phase supports interesting and important chemistry, too. There are bubbles in chemistry and there is chemistry in bubbles—in balloons and in the air we breathe.

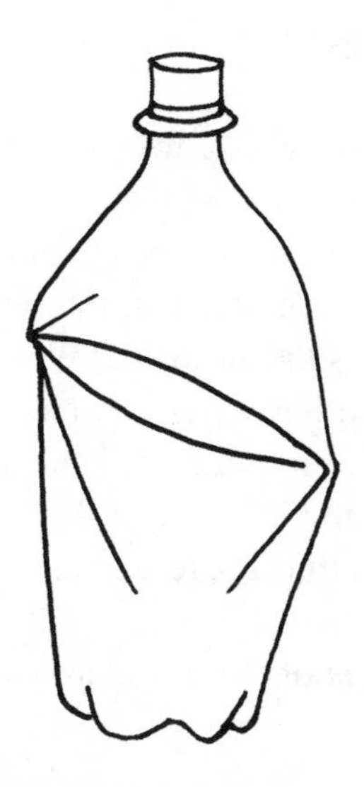

## Demonstration 10: The Soda Bottle Crunch

> *The great create an atmosphere which reacts badly upon the small. . . . It is like a chemical reagent. One day of it, like one drop of the other, will so affect and discolour the views . . . that it will thereafter remain forever dyed. A day of it to the untried mind is like opium to the untried body. A craving is set up which, if gratified, shall eternally result in dreams . . . until death and dissolution dissolve their power.*
>
> —Theodore Dreiser, *Sister Carrie*, ca. 1900

We are generally unaware of the atmosphere around us, just as fish are unaware of the sea. Nonetheless, we swim in an ocean of air and are occasionally reminded when a tree is knocked down by a storm or a tornado levels a town. This demonstration is also an effective reminder of the presence and power of the pressure of the atmosphere.

Have a sink or tub full of ice-cold water ready. Take a plastic soda bottle, fill it with hot tap water, and let the hot water sit in it for several seconds. Pour out the water and quickly screw on the cap. Put the capped bottle in the cold water and wait a few seconds. The soda bottle should implode and crumble fairly impressively.

Heating a gas will cause it to expand, and cooling a gas will reduce the pressure it exerts. Hot tap water in the bottle heats the sides of the bottle, which, in turn, heats the air inside. When a cap is tightened on the bottle after heating it with hot water, the air trapped inside is warm air. When the bottle is plunged into the cold water, the gas inside is cooled, exerting less pressure on the inside wall than the atmosphere is exerting on the outside wall. The atmosphere can then push on the wall, and the result is a crushed soda bottle.

For more on properties of the gas phase—and what goes on in soda bottles—read on.

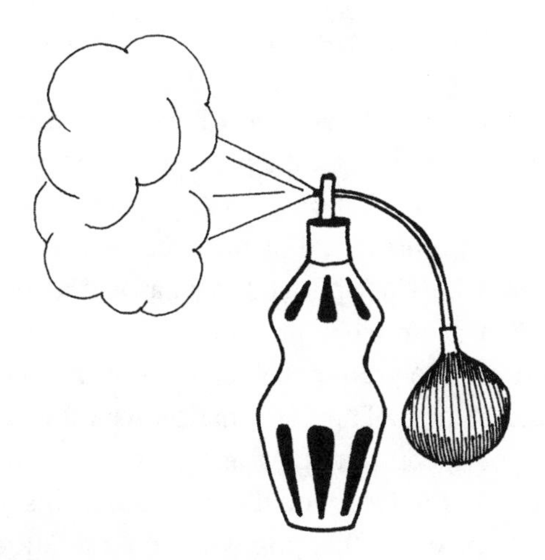

# CHAPTER 10

## It's a Gas

*(C)hemistry, Madame Lefrançois . . . the composition of manures, the fermentation of liquids, the analysis of gases and the influence of miasmata—what, I put it to you, is all this, but chemistry pure and simple?*

—Gustave Flaubert, *Madame Bovary*, 1857

Johannes van Helmont coined the word *gas* in the mid-1600s, probably derived from the word *chaos*.[1] If he did base his choice on *chaos*, it was an insightful selection, for more reasons than van Helmont knew. In van Helmont's day, most philosophers of nature thought of air as being one element, and many considered it to be a continuous fluid. Now we know that air is composed of at least three elements—nitrogen gas, oxygen gas, and argon gas—with various other compounds such as water and carbon dioxide mixed in. We also know that air is composed of molecules and atoms that are whizzing and whirling at tremendous speeds, crashing and careening. So Helmont was right about one thing. Molecules in the gas phase are caught in a continuous, crazy, chaotic dance. It is because of all this wonderful random cavorting of molecules that gases have the properties they have.

To describe the behavior of a sample of gas, we need four basic descriptors: the temperature of the gas, the pressure of the gas, the volume the gas occupies, and the amount of gas, usually given in the number of moles of whatever gases are present. For instance, if I told you I had a mole of a particular gas at 25°C (77°F) and this gas had a volume of 24.5 liters (about six gallons) at ambient pressure, you would have all the information you would need to predict the behavior of that gas should the pressure or temperature change. This information would be essential for our chemical engineer, our deep-sea diver, and even our winemaker. One of the products of fermentation is carbon dioxide, and if this gas is poorly contained or controlled, the corks that pop will be unintentional.

A gas can have considerable pressure. The pressure of the gases surrounding Earth, the atmosphere, crushed the soda bottle. The soda bottle could not be crushed when it was full, but was easily crushed when the air pressure was reduced on the inside. So why aren't we crushed when we exhale? Or for that matter why are we able to stand upright? The reason is that the pressure of the air is pushing in on us on all sides equally. If it were pushing only on the top of our heads, then yes, we would have a problem. The same principle applies to toothpaste tubes. You can squeeze out toothpaste only if you apply pressure to the middle or bottom. If you apply pressure equally to the bottom and the opening, the toothpaste stays in the tube.

The origin of the pressure exhibited by a gas, as well as the interplay among the parameters of pressure, temperature, volume, and the number of moles, is explained by a theory called the *kinetic molecular theory of gas*. The fundamental premise of this theory is that gases are made up of individual particles (hence *molecular*) and that these particles are in constant motion (hence *kinetic*, as in kinetic art). How this theory explains the properties of gases can be better understood if one compares the properties of gases to the properties of another collection of particles constantly in motion: a swarm of gnats.

Pressure is the cumulative force of molecules of a gas hitting an area, just as gnats randomly striking the walls of a container would exert a force on those walls. If the container were made smaller, then there would be an accompanying increase in the number of gnat-container collisions and an increase in gnat pressure. Likewise, all other things being equal (that is, the temperature and the amount of gas), the volume will directly influence the pressure of a sample of gas. This effect can be demonstrated with the basting syringe mentioned in the "Shopping List and Solutions." Working with the syringe without the needle attached, you can draw a sample of air up into the syringe and then put a finger over the open end where the needle is normally attached. Applying pressure to the plunger will cause the volume to decrease. At first it is easy to compress the gas, but then it becomes more difficult. This difficulty is due to the fact that pressure and volume are inversely related: as the volume goes down, the pressure goes up, a relationship known as Boyle's law.[2]

Robert Boyle, an English natural philosopher in the 1600s, did not actually discover the relationship that has become known as Boyle's law—his investigations were to confirm observations reported by others—but he recognized the significance of the findings and enthusiastically brought them to the attention of his fellow scientists. And his enthusiasm was entirely justified. With this relationship, he removed chemistry from the realm of magic and established it as a quantitative science: a science of equations and measures and laws as firm as the laws of mathematics.

In fact, all the heroes of gas-phase chemistry seem to be a rather buoyant lot. The person most often cited, at least in the English-speaking world, as being associated with another important property of gases—that of the relationship between temperature and volume—was hot-air balloonist Jacques Charles.[3] This relationship is known as Charles's law. However, Charles, who did his studies on gases at the turn of the nineteenth century, was a better balloonist than an experimentalist. From his results, he could not draw any definite conclusions about the relationship between the volume and the temperature of a gas. The reason Charles's name became associated with the law is because the person who wrote the history books was English and thought Charles sounded like a good English name. The person who correctly described the relationship between temperature and volume was Joseph Louis Gay-Lussac, who was also born in the 1700s.[4]

Gay-Lussac was a hot-air balloonist, too. In one instance, to gain altitude, he had to lighten the load of his hot-air balloon by tossing over a chair and some equipment. He later found out that bewildered peasants below had witnessed the fall of the chair from the sky and, lacking any other explanation, had been forced to credit the event to a divinity with a sense of humor. Through more earthy experiments, Gay-Lussac established that the volume of a gas will expand as the temperature goes up and that it will expand in proportion: that is, if the temperature increases, the volume will increase, or if the temperature decreases, the volume will decrease. It makes sense that a hot-air balloonist would be the one to describe this relationship because it is also the principle at work in hot-air balloons. Heated air expands, which means the same mass occupies a bigger volume, so the density goes down. Like a bubble in boiling water, the bubble of the hot-air balloon rises in the atmosphere.

It is difficult to demonstrate the expansion and contraction of gases with a convenient range of temperatures, that is, the freezing point of water to the boiling point of water. However, Gay-Lussac is sometimes credited with a companion relationship that is measurable: the fact that pressure increases with temperature. Measure the pressure in your tire in the morning before you drive to work and then after you arrive at work. The tire's increased temperature (due to friction with the road and from braking) will have caused an increase in pressure. Diesel engines work on this same principle but use volume to control temperature. Air is compressed, and the heat of the highly compressed air is used to ignite a spray of fuel.

The increased pressure of a gas when heated can also be explained in terms of the kinetic molecular theory. Energy added to a sample of gas causes the molecules to move around faster, and moving faster causes more frequent and higher energy collisions with the walls of their container, which makes the pressure go up. The gnat analogy is slightly less apt here, but then no analogy is perfect. It can be imagined that gnats would move around more in warmer weather, and, if contained, strike the sides of their container more often.

Because the behavior of gases depends on temperature, the scientists of the 1800s realized that they needed a temperature scale that did not have negative values. In calculations involving gases, such as the law of Gay-Lussac, a negative temperature could result in negative volumes, negative pressures, or other unrealistic outcomes. The question became, then, if you can't have a negative temperature, what point do you choose to be zero? Lord Kelvin, otherwise known as William Thomson, proposed a scale that had an *absolute zero*, or a temperature so low that nothing could ever be lower.[5] How would one derive such a scale? By watching how pressure and volume change with temperature and using this information to predict the point at which pressure and volume would go to zero. Using this point as a zero, the *absolute temperature scale*, also called the Kelvin temperature scale, was constructed. It turns out that absolute zero is –459.67° F (–273.15° C). Pretty cool.

The gnat analogy works well for the next gas-phase relationship, too, commonly called Avogadro's law.[6] Avogadro's law is named for the Italian physicist Amedeo Avogadro, who proposed in the early 1800s that equal volumes of gas contain an equal number of moles, if the temperature and the pressure were not changed at the same time. In other words, two swarms of gnats take up twice the space as one swarm, and so do two swarms of gas particles.

However, we wish to note that the relationships described so far may seem a bit unrealistic in that only one parameter may be changed at a time: the pressure changes, or the temperature changes, or the number of moles changes, or the volume changes. In reality, of course, this would rarely be the case. Pressure, volume, and temperature would change simultaneously, or the number of moles would increase while the temperature and pressure were changing, too. Fortunately, the various influences can be combined in one rather straightforward equation that accounts for all parameters: temperature, pressure, volume, and the number of moles. The equation that manages this balance of influences is called the *ideal gas equation*, and with this equation one can normally make a very good estimate of a missing parameter, such as volume, if one knows the temperature, pressure, and amount of gas.

$$\frac{\text{Pressure ? Volume}}{\text{Temperature ? Number of moles}} = \text{a constant value}$$

This equation states that the ratio of the pressure-volume product to the temperature-amount product remains constant, so knowledge of three of the parameters will predict the fourth. For instance, if the temperature, volume, and pressure are known, the amount of gas can be calculated.

We say that the ideal gas equation provides an estimate because this equation is "ideal" in the sense that it is idealized, not in the sense that it is the best possible equation. The idealization is that the intermolecular forces have been ignored. In truth, the intermolecular forces will come into play at some point, causing the gas to deviate from ideal behavior. Under pressure, a gas may condense. Pressure forces molecules into enough proximity that attractions exert their influence. On the other hand, if the pressure is too great, the gas may compress less than expected because repulsive forces serve to keep the molecules apart. But that's all right. Deviations from the ideal gas law serve to remind us that gases are not made up of featureless, geometric points but real materials with shape and volume and stickiness . . . and sometime even smell.

## For Example: Gutsy Gas

Given the demand for over-the-counter flatulence-reducing digestive aids, it might be interesting to note some of the properties and behaviors of the digestive gas we try to inhibit.

Flatulence is produced in the intestinal tract by the bacteria and yeast that live there. As we will explore in more detail during our discussion of biochemistry, these bacteria and yeast are essential because human bodies do not have the chemical equipment necessary to break down some of the complex sugars present in foods such as beans and peas. The bacteria and yeast, on the other hand, have the ability to break down these foods, but when they do, the side products are a mixture of gases, including carbon dioxide, hydrogen, and methane. These gases are themselves odorless, but in their travels through the intestinal tract they pick up minute impurities, including some odiferous sulfur-containing compounds. Methane is flammable, so it is true that some flatulence can be made to burn, but another myth—that burning flatulence can "flashback" into the intestines—is not. As we've noted, combustion requires oxygen, and the gases that are still in the intestine are relatively oxygen deprived. But even if flashback is removed as a danger, it is still a bad idea to attempt to ignite flatulence. The drawbacks are obvious.

More germane to the present discussion, however, is why does intestinal gas exit at all? Common experience tells us that the gas-phase is less dense than liquid and that gas bubbles will rise out of liquidlike materials. Our discussion of the properties of the gas phase reinforces this notion. But intestinal gas seems to travel contrary to the expected direction.

The reason for this backward behavior is *peristalsis*, the squeezing action the intestines make as they push matter through the intestine toward the exit. Peristalsis is stimulated by eating or drinking, which is why flatulence may occur immediately after eating even though the current meal is still sitting in the stomach in its early stages of digestion. If some gas does manage to percolate upward against the action of peristalsis, the many twists and turns in the intestines serve to inhibit its progress. When one is lying down, however, the up-down issue is negated, which is why one may experience a bout of flatulence right after waking.

It is possible to find digestive aids that reduce the amount of flatulence produced. (And we will see how they work when we revisit this topic in our discussion of biochemistry.) It is also possible to avoid excess flatulence by staying away from the foods that produce it, such as rice, pasta, potatoes, breads, beans, and peas. All this avoidance, though, is not for health's sake but for the sake of societal taboos. The odor of flatulence, which will vary depending on the type of food ingested, is deemed offensive in many cultures. The sound, both as the gas moves through the intestine and in venting, is considered vulgar. But no link has been made between ill health and excessive flatulence; in fact, the contrary may be true, so it is probably not a good idea to routinely struggle to suppress intestinal gas. After all, gas production is a sign that your intestines are working well, which is always a good thing to know.

## DEMONSTRATION 11: IT'S IN THE AIR

*What chemistry!*
*That the winds are really not infectious,*
*That this is no cheat, this transparent green-wash of the sea which is so amorous after me,*
*That it is safe to allow it to lick my naked body all over with its tongues . . .*

*The swell'd and convuls'd and congested awake to themselves in condition,*
*They pass the invigoration of the night and the chemistry of the night, and awake . . .*

*O love! O chant! solve all, fructify all with the last chemistry. . . .*

—Walt Whitman, *Leaves of Grass*, ca. 1855

We noted during our discussion of concentration that many important reactions happen in solution because solutions provide mobility and means of contact. There are, however, many reactions that occur in the gas phase or between the gas phase and the solid or solution phase. Gas molecules, too, have mobility.

This mobility of molecules is called *diffusion*, and the diffusion of gases can be readily demonstrated. Choose an area free of stray breezes. Put on safety glasses. Pour a little ammonia in a bowl, place it in one corner of the room, and then walk to the far corner of the room. If you wait a moment or two, the smell of ammonia should reach you because the gas-phase ammonia molecules are very lightweight and diffuse quickly. Insects rely heavily on diffusion. Pheromones, or chemical substances released as odors, are used as a primary method of communication among our more diminutive brethren. The male gypsy moth can detect a burst of a few hundred molecules of a sex attractant from a female gypsy moth located as far as three miles away.[1] Love is a wonderful thing.

Diffusion results from the fact that molecules that are in the gas phase are continuously in motion, though the average rate of movement will vary depending on the temperature and the mass of the molecule. The reasons are very intuitive. Higher temperatures indicate more energy is being provided to the system, so the particles travel faster, just as a baseball thrown with more energy will travel faster. Molecules of a higher mass travel more slowly than those of a lower mass, just as a bowling ball will travel more slowly than a basketball if both are kicked with the same energy. The mass dependence of diffusion can be demonstrated as follows.

Put on your safety glasses. Take two tall glasses of the same height and fill one with about two inches of vinegar and put an equal amount of household ammonia in the other. Be certain the heights of the liquids are equal.

Wet a paper towel and then add a few drops of phenol red indicator from the swimming pool test kit suggested for purchase in the "Shopping List and Solutions" onto the paper towel. Put the paper towel with the indicator spot over the mouth of the glass with vinegar. Slowly, in about three minutes, the paper towel should turn yellow. This color change is due to the acetic acid molecules escaping from the vinegar and diffusing, in the gas phase, to the indicator on the paper towel. When the towel is yellow, pick it up and place it over the glass with ammonia. The towel should now turn red, but more quickly. It turns red more quickly because ammonia is a base, and it turns red quicker because the molecules of ammonia are about 3.5 times as light as the molecules of acetic acid, which puts more molecules in the vapor and helps them to move more quickly through the air. Once the paper towel has turned red, it can be placed back over the acetic acid and it will turn yellow again, but more slowly.

Now let's further explore nuances of gas-phase reactions.

# CHAPTER 11

## When Gases Put On Airs

*Ashes denote that fire was;*
*Respect the grayest pile*
*For the departed creature's sake*
*That hovered there awhile.*

*Fire exists the first in light,*
*And then consolidates,*
*Only the chemist can disclose*
*Into what carbonates.*

—Emily Dickinson, *Poems,* third series, ca. 1850

Chemical reactions that involve gas-phase reactants, their products, or both are enormously pervasive, essential—and peculiar. Here we will catalog some proclivities of reactions involving the gas phase and illustrate them through examples that may seem surprisingly familiar.

Most of the peculiarities of gas-phase reactions stem from the fact that gas-phase molecules are, on average, relatively far apart—relative, that is, to liquid- or solid-phase collections of molecules. They also fly around at a greater speed than they would have in the liquid or solid state, and as a consequence, intermolecular forces are relatively unimportant. Because of solvent cages and viscosity resulting from intermolecular interactions, it may take solution-phase reactants longer to swim to one another, but it also takes them longer to swim apart, which can facilitate the reaction. In a liquid, reactants find themselves in the same vicinity for a longer period of time and might experience more collisions with one another. The more collisions, the greater the probability for a reaction. In the gas phase, the reactants have only one opportunity for reaction with each approach. If gas-phase molecules do happen to bump into each other, they are often as likely to fly apart as react.

So there are a number of factors working against the success of gas-phase reactions, yet several gas-phase reactions have a reputation for not only reacting but also reacting explosively—such as the reaction between gasoline fumes and oxygen. On closer examination, however, it can be seen that many of these reactions have to be initiated, or sparked, before they take off. Sparking might cause a reaction, where otherwise there would be none, because the spark results in the formation of what is known as a *radical*, that is, a chemical species with an unpaired electron.

The term *radical* may be familiar to watchers of the evening news. Political radicals may cause riots with their overreactions, and chemical radicals cause riots in the body: overabundances of radical species have been blamed for aging processes and cancer. Previously we discussed how atoms like to collect or shed electrons until they have nice filled shells. Now we can point out that these nice filled shells almost always end up with an even number of electrons. Atoms, and molecules as it turns out, like to have an even number of electrons because electrons come in pairs. Like shoes in a shoe box, they fit together as a pair, one beside the other. When something occurs to cause an unpaired electron, a radical, that species tends to be very reactive: it is looking for another electron. Free radicals in the gas phase increase the number of successful encounters significantly.

As we saw in our discussion of redox reactions, combustion reactions can travel through the gas phase quite rapidly. Solvent fumes, likewise, can be quite flammable, which is why they should be used away from any sparks or open flame and in a well-ventilated space. The liquid solvent does not have to come in contact with the flame for there to be a problem. If solvent fumes mix with oxygen in the air, then a spark may be all that is required for a reaction.

The explosions that take place in the internal combustion engine are very similar. The gasoline is sprayed as a fine mist into the cylinder, where it is mixed with air and compressed before it is sparked. The final temperature is so high that nitrogen gas and oxygen gas from the air can react and form various nitrogen- and oxygen-containing compounds such as NO and $NO_2$. Collectively, these gases are sometimes referred to as NOx (pronounced *knocks*), and the red-brown color of the NOx compounds account for the red-brown color of smog. The fumes you may have noticed when preparing copper sulfate solution as described in the "Shopping List and Solutions" are NOx compounds.

Hydrogen mixed with oxygen triggers another notoriously explosive gas-phase reaction that most probably contributed to the explosion of the space shuttle Columbia as well as the Hindenburg disaster. Gas-phase explosions usually react via chain reactions: the electron in a radical finds a mate, but in the process steals an electron from another pair, which creates at least one other radical and possibly more, if a bond is disrupted.

Because of this chain of events, a gas-phase reaction can vary in its explosive character depending on the total pressure, that is, the number of potential target molecules present. At low pressures, the hydrogen-oxygen reaction fizzles because the reactants aren't at a high enough concentration. At moderate pressures, the reaction is explosive, but, interestingly, at very high pressures the mixture becomes less explosive again. The reason for the change is the *quenching* of radicals. At very high pressures, radicals can lose energy by collisions with the walls of their container and form passive pairs rather than explosive chain reactions. At still higher pressures, there are so many reacting molecules that the walls can't defuse the radicals quickly enough to stop an explosive situation, and an explosion may take care of the question of the wall—permanently. Oxygen figures prominently in many explosive reactions because oxygen gas exists naturally as a *diradical*: two of the electrons around the oxygen molecule are unpaired, and each unpaired electron is reactive.

So far we have been discussing reactions where both reactant and product are in the gas phase, but these are not the only reactions in which the gas phase is involved. Sometimes the reactant is solid, and the product is in the gas phase. One might suppose that solid-to-gas reactions would behave rather lazily because molecules in a solid have limited mobility. However, gunpowder is an example of a solid-to-gas reaction that behaves quite vigorously when sparked. In gunpowder, solid nitrates, carbon, and sulfur turn into gaseous NOx, carbon dioxide, and sulfur oxides so rapidly that the expansion can typically project a bullet at speeds of around one thousand feet per second. Similarly, the decomposition of sodium azide, a sodium nitrogen compound, into nitrogen gas fills automobile airbags in a twentieth of a second. Sensors detect the impact and send an electric spark to an igniter reaction, which creates enough heat to start the decomposition of sodium azide into nitrogen gas.

Solution-phase reactions that produce gases include some reactions we want to avoid. It may be recalled that one of the admonitions in "A Few Necessary Words on Safety" was not to get creative with mixtures of chemicals, and one important reason, among others, is gas-phase products. Mixtures of household cleaners can produce noxious gases capable of causing difficulty in breathing or worse. Gas-phase products are particularly insidious in this regard because they spread throughout the room. Even seemingly innocent gases can cause suffocation if they are concentrated enough to displace the oxygen we need to breathe. At one point in European history, carbon dioxide was called *gas Sylvester*, Sylvester coming from the Latin for "forest." The gas could be found in caves that contained rotting wood because carbon compounds in wood decompose into carbon dioxide. Carbon dioxide gas is heavier than air, so in unventilated caves, it settled to the ground. Dogs coming into the caves would suffocate, while humans, breathing higher above ground, would survive. The whole process, of course, would be invisible and, at that point in European scientific understanding, very mysterious. For this and several other reasons, the early Europeans may be forgiven for their seemingly irrational belief in magic and superstitions. Seeing your dog drop dead from an invisible force that had no effect on you might convince anyone of the supernatural!

Reactions involving gas-phase reactants can also be difficult to coax along because the reactants have to be concentrated, or corralled, before a significant reaction can take place. Corralling gas-phase reactants generally means increasing the pressure of the particular gases. In one famous gas-phase reaction, one in which nitrogen and hydrogen react to form ammonia, the pressure of the gases is high, the reaction mixture is heated, and a facilitator, a catalyst, is used.

The process, called the *Haber process*, produces ammonia, which can be converted into nitrates, an indispensable ingredient in gunpowder. Before this process was developed, nature supplied the nitrates. Some plants, such as pea and peanut plants, harbor bacteria in their roots that are able to *fix* nitrogen from the air, that is, convert gas-phase nitrogen into compounds that can be used by the plants. Animals eat the plants, and nitrates form in their nitrogen-rich manure. Large supplies of manure, such as those found in seasonal roosting areas for large flocks of birds, used to be harvested for nitrates. In Europe prior to World War I, such nitrates had to be imported, which meant the importing countries were vulnerable to blockade. So the hunt was on for a more domestic method of manufacturing ammonia from the nitrogen gas in the air.

The person who first devised a practical process, the German chemist Fritz Haber, received much adulation, many honors (including the Nobel Prize in 1918), and significant consideration when it came to the dispensing of funds and new facilities. When war ensued, he continued in the service of the German government and agreed to work on the development of mustard gas and other chemical weapons. At this point, his life then took a downturn. His wife committed suicide, Germany lost World War I, and Haber, a Jew, was forced to flee Germany when the Nazis came to power.[2]

Haber's breakthrough was to find the catalyst, a facilitator for the reaction, which in this specific case was an iron and iron oxide solid—rust. Solid catalysts can facilitate reactions because molecules are three-dimensional beasts, and when it comes to reactions, orientation matters. This restriction can be understood by considering interactions between other three-dimensional objects: a kiss is just a kiss, but three-dimensional humans have to be oriented correctly for the kiss to be on target and effective. One advantage of a solid catalyst may be that it can hold the reactants in a favorable orientation. Automobiles use catalytic converters to convert NOx back into nitrogen and oxygen and convert poisonous carbon monoxide to carbon dioxide.

Other reactions that involve gases and catalysts are ones in which the reactants are in the gas phase but the product is a solid or liquid. The Fisher-Tropsch process is an example of such a reaction.

In the Fisher-Tropsch process, carbon monoxide is generated by treating coal with steam, a process called coal gasification.

$$C + H_2O \rightarrow CO + H_2$$

The mixture of product gases, called synthesis gas, can be enriched in hydrogen via the *water-shift reaction*,

$$CO + H_2O \rightarrow H_2 + CO_2$$

If this is done enough times, fairly pure hydrogen can be generated for use in Haber ammonia synthesis. The carbon monoxide and hydrogen can also be converted to methanol, an alcohol that can be burned as a fuel, or, if repeated many times and with a catalyst, converted into the long-chain carbon compounds that make up waxes and oils.

$$CO + 3\,H_2 \rightarrow CH_4 + H_2O \rightarrow \rightarrow \text{long chain of } CH_3CH_2CH_2CH_2 \ldots$$

So we see that interactions between solids, liquids, and gases are essential to many important industrial reactions. In addition, other reactions involving gas, solid, and liquid interactions are vital to us on a nonindustrial level. For instance, the interactions between us and the air we breathe.

## For Example: Hardwired for Respiration

Perhaps the most striking example of a reaction involving the gas phase is respiration, the process by which organisms exchange gases with their environment. In the grand symbiosis of planet Earth, plants consume carbon dioxide and release oxygen, while their animal allies consume oxygen and release carbon dioxide. Plants, however, also need oxygen, and the explanation for this involves—perhaps surprisingly—batteries.

Batteries power electric devices by providing a stream of electrons. A stream of electrons has been compared to the flow of a river (in fact, both are called *currents*), and the analogy is a good one. Just as a stream of water can do work by turning a water wheel, a stream of electrons can do work, too. A battery produces a stream of electrons by having two reactions—one that produces electrons and one that consumes electrons—physically separated from each other but able to be connected by a wire.

A battery needs two reactions because the current of electrons can only flow if it has a source *and a place to go.* The second requirement—a place to go—is very important. If the stream is dammed up, the water does not flow. You don't have a stream, you have a pond. Similarly, for electrons to flow in an electric current, they must have a place to go. Storage batteries can store charge because their electrons can't leave if they have no place to go. To produce the stream of electrons, the ends of the storage battery have to be connected by a wire, creating the streambed, if you will.

Plant processes can also be powered by streams of electrons from coupled reactions—one that provides electrons and one that consumes electrons—and these reactions can also be separated. To allow the flow of electrons, plant reactions must also be connected, but instead of a wire, plants have something called an *electron transport chain.* Here's how it works.

Plants are autotrophs, which means they are capable of making their own food, but they don't make it out of thin air. They make it out of thick air: air thick with carbon dioxide, $CO_2$. Plants take in carbon dioxide and water and produce sugars and oxygen through a magnificently orchestrated chain of events called photosynthesis. Then, in a manner similar to that of their animal counterparts, plants metabolize sugars for energy and in the process consume oxygen. Each step in the process has to be powered, and the power can come from the flow of electrons, just as a flow of electrons might power an electronic device. The source for the electrons in devices is usually batteries, and the sources for electrons in leaves are batterylike, too. Just a lot smaller.

In the first phase of photosynthesis, called the light phase, the reaction

$$2\ H_2O \rightarrow 4\ H^+ + O_2 + 4 \text{ electrons}$$

provides electrons to chlorophyll, the green pigment molecule in plants. Sunlight prompts the electrons to flow from chlorophyll into the electron transport chain, the wiring in the leaves. The use of light to excite electrons has become an everyday event with the proliferation of remote-control devices. Light from the remote control excites electrons in a semiconductor receiver located in, say, a TV. The resultant current, in turn, does work by switching on or off the electronic circuitry of the TV. In leaves, light excites electrons in chlorophyll, and the resultant current runs through the electron transport chain. The electron transport chain is a series of molecules that can pass along electrons. It is located in a plant structure called the thylakoid membrane.

As we pointed out, however, the electrons will not flow unless they have someplace to go. In the light phase of photosynthesis, the electrons moving down the electron transport chain are taken up by nicotinamide adenine dinucleotide phosphate, a molecule with nearly as many symbols in its formula as letters in its unwieldy name. Luckily, it suffices to refer to this entity by its initials, $NADP^+$, a practice common in the complicated world of biochemistry.

A salient feature of $NADP^+$ is its net positive charge, so it can act somewhat like an electron sponge.

$$NADP^+ + H^+ + 2 \text{ electrons} \rightarrow NADPH$$

In the process of flowing down the electron transport chain to $NADP^+$, the electrons do work. In this case, the work they produce is used to create a *proton gradient.*

At this point, we must interject an aside: hydrogen, as may be recalled, consists of one proton and one electron when it is a neutral atom. When its one electron has been stripped away, which is the case for $H^+$, the resulting hydrogen ion is often referred to as simply a proton because that is just what it is, a proton and nothing more. A *gradient* in science has the same connotation as a gradient on a hill or on a road, that is, a difference in size between two quantities, such as height, or in this case, the number of protons, separated by some distance. Therefore, the movement of hydrogen ions from one side of the thylakoid membrane to the other is referred to as setting up a proton gradient.

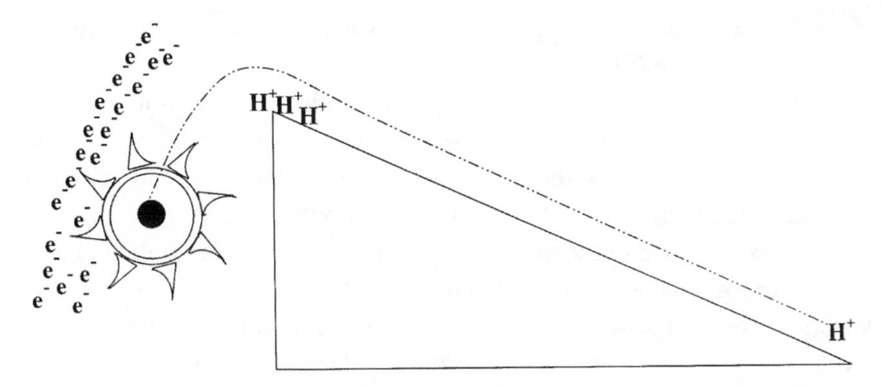


Figure 1.11.1. The work done by the flow of electrons down the electron transport chain is used to move protons to the top of a proton gradient.

Why would the leaf want to set up a proton gradient? So the protons, like the electrons, can flow from where they are in excess to where there is a deficit and in the process do work. The work that the protons do is to power the manufacture of a chemical found in all living cells: ATP.

Adenosine triphosphate, better known as ATP, is able to store energy for cellular processes. When ATP reacts, the energy stored in its bonds is used to drive various processes such as the transport of materials in the cell, the synthesis of compounds needed by the cell, and, in animals, muscular contraction. To generate enough ATP to do all this work, however, there have to be other ATP-generating mechanisms, which begin with the dark phase of photosynthesis.

The dark phase is "dark" because it does not require light. The dark phase is also called the *Calvin cycle* for the chemist who first described it, Melvin Calvin. In the dark phase, carbon, from carbon dioxide gas, is "fixed," that is, converted into a less volatile form, sugar, that can be used by plants and animals as a source of energy. Oxygen gas is produced in the overall reaction but given off to the atmosphere as a by-product:

$$6\ CO_2 + 6\ H_2O \rightarrow C_6H_{12}O_6\ \text{(glucose)} + 6\ O_2$$

One of the power sources for this reaction is the NADPH generated in photosynthesis. ATP is also involved in these reactions—in fact, used up in these reactions—but the next stage makes up for the lost ATP and more. The next step is to break down the sugars for energy in a process called *cellular respiration.*

Both plants and animals carry out the cellular respiration stage of metabolism. Although photosynthesis can occur only during the day and some plants hibernate without photosynthesis in the winter, respiration must continue, day and night, winter and summer. The cellular respiration that takes place in the plant cell is remarkably similar to the cellular respiration that takes place in the animal cell. In cellular respiration, the sugars are broken down to produce ATP. Again, as in photosynthesis, the work is done by an electron current running through an electron transport chain. Electrons flowing down the electron transport chain create a proton gradient that can be used to drive ATP formation. One of the differences is that the source of electrons is the reduced form of the molecule nicotinamide adenine dinucleotide, or NADH, not water. And this time oxygen is used to pull the electrons out at the other end, rather than $NADP^+$, with the oxygen-consuming reaction

$$O_2 + 4 \text{ electrons} + 4\ H^+ \rightarrow 2\ H_2O$$

which explains why plants also need oxygen, as promised.

However, it is still possible for some cells to sometimes produce energy without oxygen. The oxygenless energy-producing process is called *fermentation.* Though fermentation is often taken as meaning the process that produces alcohol, it actually covers several processes that operate without oxygen. For instance, when our muscle cells are oxygen deficient, such as when we exercise vigorously and our breathing cannot keep up with our oxygen demand, our cells must rely on methods for producing energy that do not require oxygen. The end product of one of these alternate, oxygenless methods is a kind of acid known as lactic acid. As lactic acid builds in oxygen-deprived muscle cells, the cells are less able to function, and the result is weakness, pain, and fatigue. Yeast relies on another type of oxygenless energy generation that has ethanol as its end product. We take advantage of this fermentation when we make wine and other alcoholic spirits.

Although oxygenless metabolism may be a pain for us, there are microorganisms that thrive in oxygen-deficient atmospheres. The bacteria that spoil canned food that is not properly sterilized flourish in the absence of oxygen, and their gaseous waste can cause cans of spoiled food to bulge. But energy generation without oxygen is not as efficient, and the life forms that rely solely on oxygenless energy are not very big. There is speculation that dinosaurs grew as large as they did because there was more oxygen in the atmosphere at that time, and there seems to be some geological evidence that oxygen levels have slowly decreased.[3] Apparently, we are now about the right size to efficiently use the present oxygen supply, but should it go down again, things might be different.

Life would go on, but the creatures would resemble us even less than we do the dinosaurs, and less than the dinosaurs resemble the single cells from whence they came. But those single cells were responsible for producing the oxygen in the atmosphere that eventually accumulated to the point that allowed multicellular life, so it might be that this cycle would eventually repeat. We are used to thinking linearly, in terms of a beginning and an end, because that is how we reason. But nature is not so constrained. Nature tends to appreciate cycles, and life has its ups and downs.

# Demonstration 12: How Does Your Garden Grow?

*I feel like a white granular mass of amorphous crystals.*
—Lafcadio Hearn, quoted in
*The Life and Times of Lafcadio Hearn*, ca. 1900

Gas and liquid phases are not the only forms of matter with reactions of interest to chemists. The solid phase, though seemingly fairly inanimate, can have a lively side, too.

Put on your safety glasses. Select a small glass or plastic bowl similar to a cereal bowl to serve as your garden, making sure it is a bowl you are willing to discard when the demonstration is over. Place this bowl inside a larger bowl, box, or container or on a large sheet of newspaper or aluminum foil, so that any crystals that fall from the garden will remain inside and make clean up easier. Find a location where your garden can sit undisturbed and unmolested and where the smell of ammonia will not bother anyone.

Take a small kitchen sponge, wet it, and wring it out so that it's just damp. Cut a strip of the damp sponge into little cubes and place the sponge cubes into the garden. Spread them evenly in the bottom, about one-half inch (1 centimeter) away from each other and the sides of the bowl. The size and number of the cubes will depend on the size of the bowl. It will probably take three to five cubes. You are now ready to add the materials that will form your garden.

Take the laundry bluing, household ammonia, table salt, and distilled water that were suggested for purchase in the "Shopping List and Solutions" and get a plastic teaspoon. Add two teaspoons each of bluing, distilled water, and ammonia to the bowl. If you want, you can add color by dropping a few drops of food coloring directly on the cubes of sponge. Precise amounts are not crucial to the outcome. Swirl the bowl slightly. Sprinkle two teaspoons of table salt over the cubes. Don't worry if some salt gets into the solution on the bottom of the bowl. Let the bowl remain undisturbed overnight. The next day, sprinkle two more teaspoons of table salt over the cubes and let the bowl sit once more overnight. On day three you should repeat the addition of all materials from day one, taking care not to disturb any crystal growth that has occurred. The salts will blossom into profuse, delicate, fingerlike crystals. Looking at the crystals through a magnifying glass will further enhance the garden tour.

# CHAPTER 12

## Crystal Clear Chemistry

*The road is wide and straight and bright as crystal, and the sun is at the end of it.*
—Feodor Dostoyevsky, *Brothers Karamazov*, ca. 1880

The word *chemical* can conjure up images of goo, globs, and glops, but usually not pictures of nice, neat, orderly structures. But solid-state chemicals can be wonderfully symmetric and elegantly structured. Beautiful, mathematical symmetry is evident in the crystal garden demonstration and is exhibited in even a grain of salt. Here we will examine the forces that bring about such structure and other phenomena associated with the solid state of matter.

Thus far, we have not dwelled on the three-dimensional nature of molecules because this added layer of complexity was not needed. Historically, this was the order of events, too—more or less. It wasn't until the mid-1800s, a half century after Dalton's theory of the atoms became generally accepted, that a substantial number of European scientists began to grapple with the possibility, and the consequences, of three-dimensional arrangements of atoms in molecules.

One of the first pieces of evidence that there might be something fundamentally important in the way the atoms arrange themselves in compounds came from the discovery of *isomers*, molecules with the same chemical formula but with different properties. As discussed earlier, both grain alcohol and methyl ether have the same molecular formula, $C_2H_6O$, but one is a liquid at room temperature, and the other is a gas. The odors are also distinctively different, though it is not recommended to inhale either one. In certain concentrations fumes of methyl ether or grain alcohol can kill. Glycogen and cellulose are both long chains of glucose molecules linked together, but while glycogen stores energy in the human body, the bonding in the cellulose isomer is such that the human digestive system cannot break it down into food. During the Irish potato famine of the mid-1800s, people were reduced to eating grass, but they starved anyway, because they could not metabolize the glucose chains found in the grass.

In the 1800s, even when it was recognized that two compounds might have the same formula but different arrangement of the atoms, it took a while before it was generally acknowledged that the three-dimensional angles and spacings between the atoms could influence their behavior. Louis Pasteur, savior of the French wine industry in the 1800s, found that a salt of tartaric acid, found in wine, existed in two distinct forms.[1] Crystals of the two forms could, with care, be separated by tweezers if viewed through a magnifying glass. Because the crystals were both tartaric acid, the elements in the molecules were bonded in the same order, so the difference had to be the way the bonds were arranged in space.

When this logical explanation was first put forward, however, it was not met with instant universal approbation. The proponents of the idea were labeled daft, or worse, and had to endure the criticism. But time proved them out. To explain the observed crystal structure in salt, sugar, and the crystal garden, we now embrace the three-dimensional nature of atoms, molecules, and ions in our three-dimensional world.

To understand the forces behind the three-dimensional structure of molecules, we need to refer back to an earlier analogy. We said that the nucleus of the atom, being made up of protons and neutrons, is thousands of times as massive than the electrons that surround it. The electrons, we continued, are like fleas on an elephant. But though the fleas are small, they certainly influence the behavior of the elephant, and three-dimensional structure of molecules is one place where this influence is felt. The three-dimensional structure of crystals comes about because of one basic fact: electrons like to be paired, but the pairing is strictly monogamous. Once paired, electrons hold all other pairs at a polite distance. In the parlance of the chemist, this is called *electron pair repulsion.* To see how this plays out in molecules, consider water.

As you may recall from our discussion of intermolecular forces, the water molecule has a **V** shape in which the oxygen is at the point of the **V** and a hydrogen sits at each tip. At the time we did not say why water assumes this shape, but we can address that question now. As can be discerned from the periodic table, shown in figure 1.12.1, oxygen has eight electrons around it, but two of these are in the first closed shell. The outer six, the valence electrons, are the ones that really influence the chemical behavior of the atom. Oxygen will be happiest, it may be recalled, when it has acquired two more electrons to fill its outer shell. Hydrogen has one electron and is looking for one more to fill its first shell, as can be determined by again consulting the periodic table, shown in figure 1.12.1.

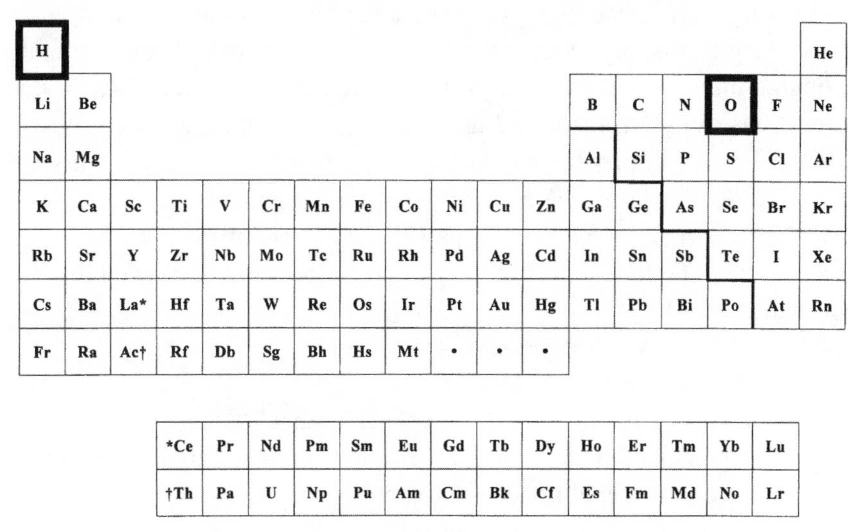


Figure 1.12.1. Hydrogen, situated at the beginning of the first row, needs one more electron to fill its shell. Oxygen, situated two from the end of the second row, needs two more electrons to complete its shell.

Chemists use a visual aid called *Lewis dot structures* (named after the United States chemist G. N. Lewis) to show how nuclei in molecules arrange themselves so that as many nuclei as possible are surrounded by filled shells of electrons.[2] The device starts by putting a dot for each valence electron (hence Lewis *dot* structure) around the symbol for the elements. Hydrogen and oxygen might be represented as follows:

H· ·Ö·

In this Lewis dot structure, hydrogen has one dot for its one electron; oxygen has six dots for its six valence electrons.

Displayed in this manner, it is easier to see how two hydrogens might combine with one oxygen to make water. They can share electrons so that each hydrogen has a filled shell of two electrons and oxygen has a filled shell of eight electrons.

H:Ö:H

However, this arrangement still does not look a lot like a **V**. To explain the **V**, it is necessary to evoke electron pair repulsion, as well as the occupancies of the valence shells, in a key theory called, appropriately enough, *valence shell electron pair repulsion* theory, or VSEPR. In this theory, electrons are identified as *bonded pairs*, which are electrons in bonds, or as *lone pairs* (or *unbonded pairs*), which are electron pairs not in bonds. For instance, the bonded pairs are the pairs of electrons being shared between oxygen and hydrogen in the Lewis dot structure for water given above. The lone pairs are the pairs on either side of oxygen that are not being shared.

The tenets of VSEPR are straightforward:

- lone pairs (unbonded pairs) repel each other
- lone pairs (unbonded pairs) repel bonded pairs, only less so; and
- bonded pairs repel bonded pairs, but even less so.

Understanding these principles allows us to analyze the shape of the water molecule as follows.

All the electron pairs, those in bonds and those in lone pairs, want to get as far away from each other as possible, but the two lone pairs on oxygen want this the most. So the lone pairs spread apart and, in this action, force the bonded hydrogens closer together, as is shown in figure 1.12.2.

But figure 1.12.2 still looks only two dimensional because we are forced to draw it on two-dimensional paper. To visualize the situation in three dimensions, try using balloons. Choose balloons of the same shape and size but two different colors, such as red and green. Blow up two of the red balloons and two green balloons to the same volume and tie their ends all together. (This is easier to do if you don't overinflate the balloons.) You should end up with a three-dimensional cluster in which the balloons are forced out in four separate directions. The three-dimensional shape you have created is called a *tetrahedron.*

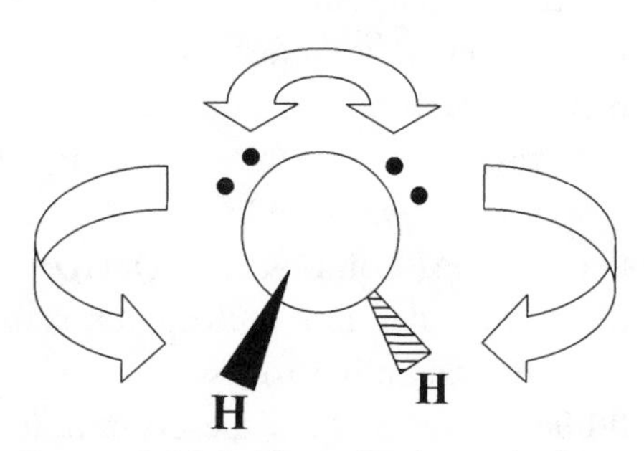

Figure 1.12.2. The electrons in lone pairs repulse each other more than the electrons in bonded pairs, so the lone pairs spread out and the hydrogens are squeezed together, which gives water its V shape.

Now blow up four more balloons, but make the red balloons bigger than the green ones (but keep each red balloon the same size as the other and each green balloon the same as the other). When you tie these together as before, you have a situation that is more representative of water: the larger balloons will force themselves farther away from each other and squeeze the smaller ones together. Taken together, the three-dimensional shape of the molecules determined by VSEPR and the trade-off between ionic attractions, covalent bonding, and intermolecular forces account for the three-dimensional structure of crystals.

Some crystals are *ionic*, that is, long networks of ions held together by ionic attractions. Sodium chloride, table salt, is an ionic solid. In crystals of sodium chloride, a sodium ion is surrounded by six chlorine ions—one on top, one on the bottom, and one at each compass point—and each chloride ion is surrounded by six sodium ions in the same manner. Crystals of sugar, on the other hand, are held together by intermolecular attractions. They fit together for optimum balance of attraction and repulsion so that a quite orderly crystalline structure results. Pure metals are stacks of identical atoms so the bonding between one pair cannot be any different than the bonding between the next pair. Consequently, it is impossible to say that the outermost bonding electrons belong to any one atom, and this independence is reflected in the excellent electrical conductivity of metals: when hooked up to a power source, such as a battery, electrons will travel freely from one atom to another in a current.

Covalently bonded solids such as quartz, diamond, and graphite form another class of crystals. Quartz is a continuous network of silicon dioxide bonded in a uniform, crystalline arrangement. Sand is a mixture of quartz and other rocks. Glass is solid quartz that has melted and resolidified without the same crystalline uniformity, in the way that melted butter does not re-form the same type of solid when it cools. Glass has been known to form naturally in lightning strikes on sand.

Diamond is a network of carbon atoms, each carbon entity bonded to the next through a covalent bond in a tetrahedron arrangement, as shown in figure 1.12.3. Diamond's sibling, solid graphite, is also made of pure carbon, but in graphite the carbons are bonded in sheets that can slide over one another, as shown in figure 1.12.4. The difference in properties—and in price—is considerable. Graphite is actually the more stable form of solid carbon, as we will discuss shortly, but the length of time required for the transformation is about as long as the age of planet Earth.

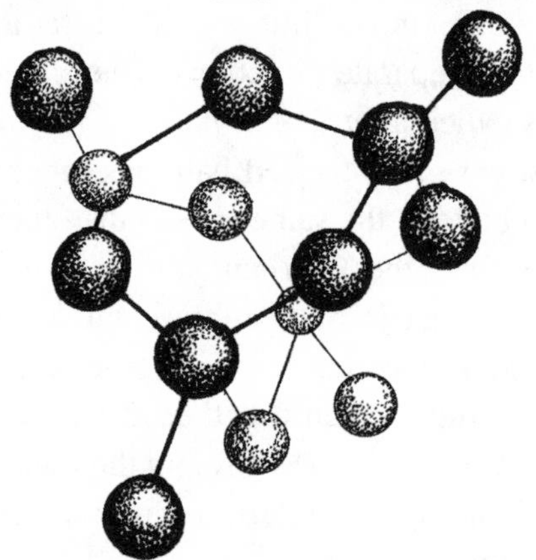

Figure 1.12.3. Diamond is a covalently bonded network of carbon atoms that assume a tetrahedron shape at each carbon.

An interesting sort of hybrid of ionic and covalent molecules can be found in salts of *polyatomic ions*. Polyatomic ions are charged groups that contain several different types of atomic nuclei—such as $CO_3^{2-}$, an ion that is familiar to us from our baking soda demonstrations. Baking soda is sodium bicarbonate, which is $NaHCO_3$. Another polyatomic ion that we have dealt with, although not explicitly until now, is the sulfate ion, $SO_4^{2-}$. As may be recalled, the superscript 2–, read "two minus," indicates that the ion has a minus two charge. Copper sulfate, the compound that forms the lovely blue-colored solutions we have used in several demonstrations,

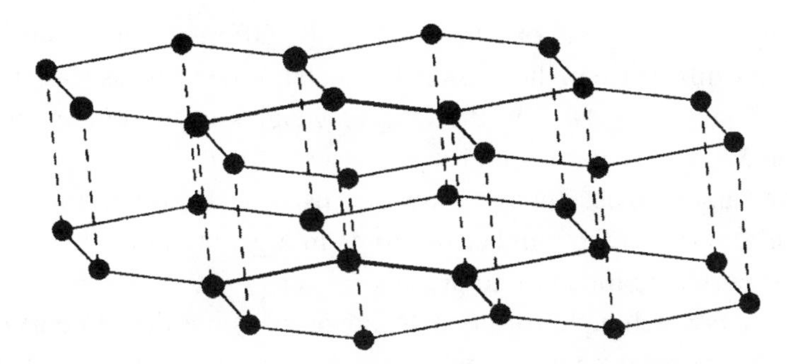

Figure 1.12.4. Graphite is covalently bonded carbon in hexagonal sheets.

is a salt of the copper ion and the sulfate ion. The bonding is covalent within the carbonate and the sulfate ions, but between the positive ion (sodium or copper) and the polyatomic ion, the bonding is ionic. These crystalline salts of polyatomic ions were used in the mid-1800s to answer an age-old question: How do you weigh an atom?

The question is obviously a very difficult one, but it has always been a very important one, too. Its significance arises from the fact that atoms react one to one. So if you want to do any kind of accounting for their actions you are going to have to be able to count them—but they are so darned tiny that you cannot. What you can do is weigh a big collection of atoms, such as a spoonful or a cupful, and calculate how many atoms are in the spoonful or cupful, but in order to know how many atoms you have in that spoonful or cupful, you have to know the mass of one atom. Scientists began to get a handle on this when they realized that there are many different metals that form sulfate salts, and all of these crystallize in approximately the same way. How does this help? It has to do with density.

Density, as you may recall, is the amount of mass of a substance per unit volume. Density varies from substance to substance; for instance, the density of a bag of feathers is considerably different from that same bag filled with bricks. As a characteristic, density is very useful because you don't need a specific amount to measure it. The density of bricks is going to be the same if you are looking at one brick or a brick wall. A spoonful of table salt will have the same density as a cupful of table salt because density does not depend on sample size. To measure the density, you need only to measure the mass of a specific volume.

For sulfate crystals, it was assumed that each "brick" of crystals contained the same number of sulfate units and metal units whether the metal was copper, magnesium, or whatever. So any difference in density had to be due to a difference in the mass of the metal. When a mass was assigned to one of these metals, then the relative weights of the others could be determined. The idea was clever and useful for its time, but has been improved upon since. Today, x-rays are used to determine the arrangement of atoms and the number of atoms in a given volume of a crystal, and the results are much more precise.

The x-ray technique for crystals, however, is not the same as that of a medical or dental x-ray. In the case of a medical or dental x-ray, the x-rays are allowed to pass through the body and the images are of the shadow cast by the dense tissue—bones or teeth—that the x-rays could not penetrate. X-ray studies of crystals (called x-ray *crystallography*) look at reflected x-rays, rather than transmitted x-rays, and analyze the patterns in the reflected rays.

Crystals reflect x-rays somewhat as CDs reflect visible light. The visible light bounces off the grooves in a CD in such a way that the light separates into its various colors, giving the CD a rainbow cast. The various layers in a three-dimensional crystal act somewhat like a miniscule groove pattern in a CD. X-rays bounce off microscopic crystal structures and produce a pattern on an x-ray film. From the particulars of the pattern it is possible to discern the size and structure of a single unit of the crystal. The position of the pieces—atoms, ions, or molecules—that make up the crystal can be extracted from this information. Once the volume of the smallest unit of a crystal and the number of atoms that it holds is known, the mass of just one piece—an atom or a molecule—can be determined from the density.

Of course, all that glitters is not gold, and not all solids are crystalline. The solids that are not crystalline are generally referred to as amorphous. These solids include rubber, some plastics, waxes, and glass, though there may be some debate over the actual assignment. The label, amorphous solid, is a human attempt at classification that nature does not have to obey. Amorphous solids can be distinguished from crystalline solids in that amorphous solids do not have a clear, distinct melting point. Crystalline solids exhibit very sharp melting points if they are pure, and therefore, a sharp, abrupt melting point is used as evidence of purity.

Water is a most interesting solid for a couple of reasons. First, it is one of the few solids that expands when it freezes rather than contracts. This expansion explains why ice floats in our drinks. The expanded solid water is less dense than the liquid water—it has fewer molecules per unit volume. The reason for the expansion is hydrogen bonding. When water freezes, it is happiest when its oxygen aligns with neighboring hydrogens, sort of like Tinker toys. Tinker toys are wooden sticks that can be joined together by connectors and made into various structures (windmills and houses are some universal favorites). Taken apart, as pieces lying in the box, the pieces don't take up that much space. But in a rigid structure, connected together, they take up far more space. Because of its **V** shape and need to align for optimum hydrogen bonding, water solidifies with large gaps that are normally filled in the freely moving liquid state. (See figure 1.12.5.)

Because of the gaps, water expands when it freezes. The expansion of water as it freezes is responsible for the cracking of sidewalks in the winter and is part of nature's plan for breaking up rocks into soil. Water is also intriguing insofar as it can satisfy the demand for hydrogen bonding with several different crystalline shapes. One of the most famous is the hexagonal structure that gives rise to the six-sided symmetry of snowflakes.

Snowflakes develop when new water molecules strike the surface of a seed crystal. They grow into six-armed structures because the molecules that strike at the tips of the crystal are more likely to stick: the point projects into space and snags molecules floating by, just as debris catches on a branch jutting into a stream.

In general, surfaces are different creatures than the bulk that spawns them. For instance, a water drop forms itself into a sphere because the water molecules inside the water drop have all their opportunities of hydrogen bonding met and are content. The water molecules at the surface have some unhappy hydrogen ends without their oxygen mate on an adjacent molecule and vice versa. This situation creates a tension, called *surface tension*, and the natural response of the system to minimize this tension

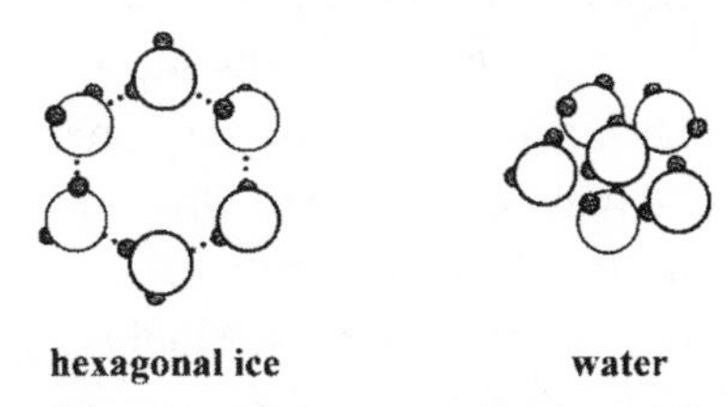


Figure 1.12.5. Hexagonal ice is the foundation on which six-sided snowflakes are built.

is to form itself into a sphere. A sphere presents the minimum surface area and thereby minimizes the number of molecules exposed to air and thus the surface tension. Solids aren't able to roll themselves up into balls, so the unique situation at the surface of solids is usually not so dramatically expressed, but the surfaces of solids are also different than the bulk material and have a special reactivity nonetheless.

The two most common concerns with surfaces in everyday experience are making things stick to surfaces when you want them to and getting them to unstick when you don't want them to. To get materials to stick to surfaces, bonding agents, or glues, are normally used. And these wonderful materials, without which we would not have essentials as diverse as plywood or sticky memo notes, have a chemistry that dates back to antiquity. Glues were originally made from animal proteins, which means they really did send old horses to the glue factory. The word *glue* has ties to the words *gluten*, which is the gooey mess derived from wheat and flour that makes dough sticky enough to form balls, and *gluteus*, which is the ball of muscle on which you sit. The ability of a particular glue to stick may be due to chemical bonding, intermolecular attraction, or both. Depending on the origin of the material—animal, mineral, or vegetable—some glues work better than others because of the type of bonding or attraction being exploited. Sometimes it is desirable to reverse the effects of glue, or unglue things, in which case it is a good idea to use a material that will insinuate itself between the glue and the glued, to break the surface bonds. Oil, such as cooking oil, is often useful in this regard. Oil can be used to remove chewing gum from most surfaces because oil can form a film on the surfaces and in effect alter the bonding properties of the surface. Because of these surface interactions, oil is related to a class of materials called *surfactants*.

A commonly used surfactant is soap. Soap has a long carbon chain that is attracted to organic material, such as dirt and grease, and an ionic end that is attracted to water. Soap can interfere with the surface attractions and is useful for unsticking things such as wedged on wedding rings or corks stuck in bottles. Machine part lubricants are also surfactants because the lubricants interfere with the attractions between bare metal parts.

Porous surfaces present a different problem because the actual surface area, including the interior of the pores, is many multiples of the apparent surface area. Therefore, a porous surface is much more difficult to clean and may resist heroic efforts by surfactant detergents. This intractability of porous surfaces is why natural materials such as wood, bone, and teeth are so prone to staining. Recent methods for removing stains from teeth involve using surfactants to hold a mild bleaching agent against the teeth. Because bones, teeth, and ivory are similar in composition, these same teeth-whitening products have been used to clean aged, stained (and, thankfully, irreplaceable) natural ivory piano keys.

The special properties of surfaces also account for the action of solid-state catalysts. The action of a catalyst often depends on its ability to hold molecules in a particular orientation that will facilitate the reaction. In the gas or liquid phase, a collision at just the right orientation is a function of chance. If the target molecule is held in a favorable orientation on a surface, the chances of a productive encounter are improved.

A solid-state *substrate* is a type of catalyst that allows many reactions or layers of product to grow on its surface. In our discussion of gas-phase reactions we mentioned that a solid catalyst could be used to grow long-chain organic molecules for waxes and oils. Substrates have also been used to grow crystalline products, such as diamonds.

## For Example: Diamonds Are (Almost) Forever

Over the eons there have been many materials, from sea shells to spices, that have been used as a medium of exchange, but the most popular have been materials that are durable and gain as much value from their scarcity as their utility. Gold is an example of the latter, as well as gemstones, in particular, diamonds. The durability of gold, its resistance to corrosion, has contributed to its usefulness for applications such as dental prostheses and electrical contacts. Likewise, the durability of diamonds has con-

tributed to the usefulness of this particular mineral in cutting tools, drill bits, and abrasives.

Demand for gold during the European Middle Ages led to many curious and creative attempts to manufacture gold from more common metals such as lead or copper. As noted, this enterprise, called alchemy, did not succeed, but fostered enough new knowledge of chemical reactions to stand as an important phase in the history of chemistry. To date, the only way to create gold artificially is via a nuclear reaction, but the process is many times more expensive than the gold it produces. Since the mid-1900s, there have been, however, methods for producing synthetic diamonds that are economically viable.

One method involves the microwave-induced removal of hydrogen from methane ($CH_4$, swamp gas) in a very sparse gas phase so that carbon atoms, stripped of most of their hydrogen, can settle out on the substrate and start building diamond crystals. In the past, the diamonds formed in this process were tiny and only suitable for industrial applications, but lately gem-quality crystals have been grown. The artificial diamond is virtually indistinguishable from the natural diamond because they are both just a crystalline form of carbon.

If diamonds are manufactured rather than mined, will that dim their sparkle in lovers' eyes? Probably somewhat. But it is hoped that someday semiconductor materials with important properties can be routinely made from the less expensive and more perfectly pure synthetic diamonds. If diamonds find use in semiconductor devices, this could give a whole new meaning to "a diamond chip."

The interest in diamonds, artificial and real, is due to their unique properties, one of which is their extreme durability. But are diamonds indestructible? Not at all. As may be recalled from our discussion of Lavoisier's proof of the conservation of mass, diamonds will burn in oxygen, if heated enough. As we shall see next, in science as in social situations, strange things can happen when matters heat up.

# Demonstration 13: Hot Packs and Cold Packs

> *The fields of Nature long prepared and fallow—the silent, cyclic chemistry;*
> *The slow and steady ages plodding—the unoccupied surface ripening—the rich ores forming beneath.*
>
> —Walt Whitman, *Leaves of Grass*, ca. 1855

Citric acid undergoes many chemical and biochemical reactions, and it also displays a property that holds particular interest for us. When citric acid dissolves in water the solution cools. Citric acid requires thermal energy to dissolve; and it draws this energy from its surroundings. A process that requires thermal energy to proceed is called *endothermic*.

To demonstrate this effect, put on your safety glasses then take a plastic sandwich bag and put in about two teaspoons (10 milliliters) of the sour salt (citric acid) suggested for purchase in the "Shopping List and Solutions." Then put one tablespoon (15 milliliters) of water into the bag and seal it up. The bag will feel cold to the touch as the sour salt dissolves. The effect is most pronounced at the beginning of the dissolution. To give yourself a good comparison, take a second bag and place just a tablespoon of water into it. The bag containing the sour salt will be clearly colder. If you have difficulty locating sour salt, try using baking soda. The cooling is not so dramatic, but still quite noticeable. Citric acid is used in some novelty self-cooling drink mixes.

There are also many processes that release thermal energy. One such reaction involves dissolving lye in water. Be aware that this process releases a great deal of heat, so you should use a lot of water with only a few crystals of lye. Also be certain your safety glasses are in place. Fill a glass that holds at least two cups (480 milliliters) with one cup (240 milliliters) of water. Don't use the bag for this demonstration because the process releases heat, which, you will recall from our discussion of the gas phase, will cause the air trapped in the bag to expand. If the reaction were in a closed bag, then the buildup of gas pressure could cause the bag to spring a leak.

Extract one half a plastic spoonful of lye from its container and introduce these crystals slowly into the glass of water. Swirl the contents slightly. The quantity of thermal energy transferred to the glass container by the process creates a warmth in the glass that can be sensed by touching the outside of the glass. Again, the effect is greatest at the beginning of the process. A process that generates thermal energy is called an *exothermic* process.

Hot packs and cold packs are sold in various sports supply stores and pharmacies for the treatment of muscle ache and sports injuries. Although usually made with different salts, the basic principles of operation are the same as the preceding demonstration.

# CHAPTER 13

## When Matters Heat Up

*You are well aware that chemical preparations exist . . . by means of which it is possible to write on . . . paper . . . so that the characters shall become visible only when subjected to the action of fire. Zaffe, digested in aqua regia, and diluted with four times its weight of water, is sometimes employed; a green tint results. The regulus of cobalt, dissolved in spirit of nitre, gives a red. These colors disappear at longer or shorter intervals after the material written on cools, but again become apparent upon the re-application of heat. . . . I now scrutinized the death's-head with care. Its outer edges—the edges of the drawing nearest the edge of the vellum—were far more distinct than the others. It was clear that the action of the caloric had been imperfect or unequal.*

—Edgar Allan Poe, *Gold-Bug*, ca. 1840

*The bit of practical teaching he afterwards reviewed with most curiosity was the course in Chemistry, which taught him a number of theories that befogged his mind for a lifetime.*

—Henry Adams, *The Education of Henry Adams*, 1920

The quote from Poe is included because it refers to *caloric*, the name that was given to heat when it was still thought to be a substance, and the quote also alludes to "the action of fire." The second citation is included because *thermodynamics*, which concerns itself with the understanding and application of the action of fire, has a bad reputation for befogging the mind. But understanding it is well worth the effort: thermodynamics embodies a powerful theory that enables chemists to predict whether a chemical reaction will take place, the extent to which it will take place, and the amount of energy that will be generated or consumed in the process—all before mixing chemicals together. Because some reactions release enough energy to weld metals, it is a good idea to know how much energy to expect in advance of experimentation. Other reactions may look good on paper but react only to a limited extent in the lab—and this information, too, is good to know before investing too heavily in a project.

Given the predictive power of thermodynamics, the theory holds a prominent place in the study of chemistry. The machinery of thermodynamics really boils down to three concepts: energy, entropy, and free energy. We will start with energy and build from there.

Our understanding of the energy of molecules is based on a molecular model of matter that assumes that all matter is in constant motion. This idea may seem a bit implausible at first, given that there are ample observations in our everyday experience that tell us some materials are solid, stable, and immobile. It is a bit difficult to acknowledge that at the molecular level a cement sidewalk is actually wriggling around like a can of anxious worms, but it is. The model of molecules in motion works well for explaining why gases and liquids behave as they do and still allows cement to behave the way it does.

Actually, the model works well only if we admit to at least three different modes of motion. These modes of motion are *rotation*, *vibration*, and *translation*. Translational motion is straight-line motion, such as walking across the room. Vibrational motion is the molecular equivalent of waving your arms up and down. Rotational motion is the molecular equivalent of spinning around in place, perhaps as an ice skater might. Vibrational and rotational motions are only available to molecules, not atoms, because the rotations and vibrations are with respect to some center defined by the bonded elements.

For instance, water has both vibrational and rotational modes. One vibrational mode is the scissor movement of the hydrogens toward and away from each other. A rotational mode is the spinning of the hydrogens about an imaginary axis through oxygen. Both these modes of motion are shown in figure 1.13.1.

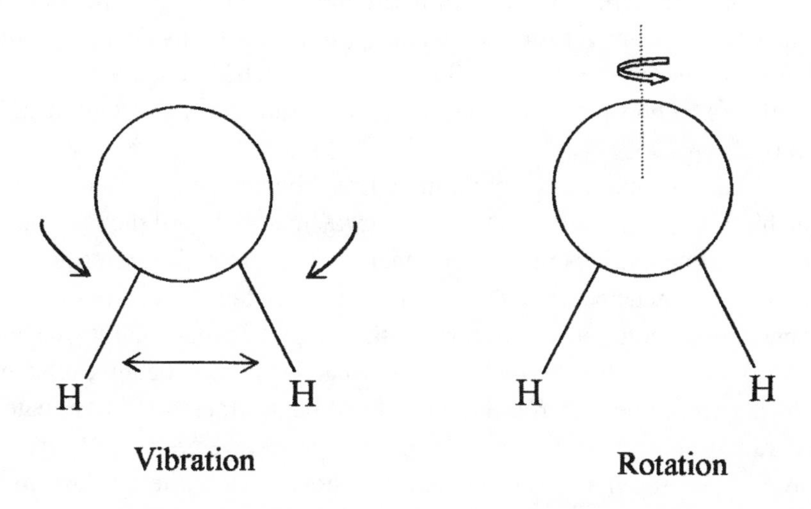


Figure 1.13.1. Water molecules are in constant motion. They vibrate, rotate, and move in a straight-line motion called translational motion. Shown here are a vibrational and a rotational mode of motion for water.

The significance of these modes of motion is that energy added to a collection of molecules will distribute itself between the various modes. An analogy might be made to a jellyfish. Poking a jellyfish will make it not only move forward but also wobble and roll about, too. In a given sample, the sum of the average energy expressed in each mode is called the *internal energy*. Every bit of matter has some internal energy, even if it is very cold. Recall that the Kelvin temperature scale defines a point called *absolute zero*, which is also the point of zero energy. But this point is merely a theoretical limit and in actuality can never be reached.

Temperature is a measure of the internal energy of an object. Using temperature as a gauge of internal energy, however, assumes that the energy is uniformly distributed among the various modes of motion. This assumption is necessary because temperature is really only the measure of one mode of motion: translational motion. A thermometer registers the energy of motion when molecules strike its surface. There are situations in which added energy does not add equally to all modes of motion, and in these cases it is possible to add energy without changing the temperature. Boiling water is a case in point. Energy is added, but the energy goes into breaking the molecules free of their intermolecular attractions so that they can go into the gas phase. A thermometer inserted into a pot of room-temperature water registers a change in temperature as energy is added from the burner—until the boiling point is reached. Then added energy continues to boil the water, but the temperature stays constant at 212°F (100°C).

When two materials are put in thermal contact, energy will flow from the high-energy material to the low-energy material until they are both at the same energy. It is the flow of energy that is technically termed *heat*, though in casual speech, the term can be used rather loosely. For instance, when we are cold, we ask to turn up the heat when technically we should ask to turn up the thermal energy. Heat is energy being exchanged, flowing from one material into another or being transferred to a material via radiation such as microwaves or sunlight. So we are correct when we say we feel the heat from the sun or a hot water bottle because this is energy being transferred to our skin.

Work is also a form of energy transfer. When you do work on a system you give that system energy at the expense of your own energy. Work can come in many different forms: you can squeeze on the system (pumping a bicycle pump can cause its temperature to rise); you can run an electric current through the system (computers have to be cooled so that their temperatures don't rise too much); and you can also use chemistry to do work. Some examples of using chemistry to do work might be burning gasoline to power a car or digesting the chemicals we call food to do all the work of our bodies.

When we use chemistry to do work or transfer energy, where does the energy come from? It comes from the breaking and forming of chemical bonds.

A bond can start to form as two atoms come together. If conditions are favorable for bonding—conditions that include the desire of each atom for additional electrons or the ability to provide electrons—electrons can rearrange into a molecular orbital instead of individual atomic orbitals. When this happens, the energy of the system is lowered. A pretty fitting analogy can be made with magnets. If the opposite poles of two magnets come close together, it requires energy to keep them apart. It is a less demanding energy situation to go ahead and let them come together. So it is with atoms that want to bond. It is a lower energy situation when they come together.

Bonding can lower the energy of two atoms, but it does not reduce the energy to zero. Each nucleus may be attracted to the electrons of the atom, but the two nuclei are repulsed by each other because each nucleus has a positive charge. So the analogy with opposite poles of magnets fails at this point. The two nuclei may settle into a bonding situation, but there is always a tension between them. When a more favorable situation presents itself, the bond can break and a new bond can form. Sometimes there is a net energy release and sometimes there is a net energy requirement. When energy is released, the reaction is termed *exothermic*. When energy is required, the reaction is termed *endothermic*.

Let's take a moment to recap. Basically, we're saying that the internal energy of a collection of molecules or atoms is the sum of their (1) rotational, (2) translational, and (3) vibrational energy. The amount of internal energy that a system contains can be changed by heat. Heat can be added to or subtracted from a system in many ways, but the method of primary interest to us is by chemical reaction. When a chemical reaction gives off heat, it is termed exothermic. When a chemical reaction absorbs heat, it is termed endothermic.

Now we add an additional item of information: molecules and atoms are lazy. Atoms and molecules like to lower their energy whenever they can. Bowling balls roll downhill, boats float downstream, and molecules settle in the lowest energy arrangement that they can manage.

But there is a catch: molecules like to be in the lowest energy configuration they can manage, but this may not be the lowest energy configuration *possible*. In the demonstration at the beginning of this chapter, we illustrated two different types of reactions, ones that give off heat and ones that absorb heat. When a chemical system goes to its lowest energy configuration, it gives off heat. Only when it goes to a higher energy configuration is energy required and only when energy is required does the system absorb heat from its surroundings. But the reaction that absorbed heat occurred freely, on its own, without any incentive other than mixing. Now we have a question to pose: Why? Do chemical systems obey different laws than bowling balls and boats? Are chemicals free to hop up hills and swim upstream?

The question is not a simple one, and indeed, there was a good deal of discussion on the nature of heat throughout the 1700s, the 1800s, and into the opening days of the 1900s. The answer has to do with something called *entropy*. Entropy is the tendency of systems to move toward a state of maximum disorder.

Drinks mix; odors drift; puddles merge: entropy is the natural tendency toward disorder. Entropy has also been called the tax that nature charges. Entropy has also been called the universal death sentence. But entropy is not just a philosophical construct or a pessimistic outlook. It is a real driver for physical processes and does indeed weigh in as a determinant in the fate of the universe.

All other considerations being equal, a system will move toward the state that has the maximum disorder. Molecules and atoms are not only lazy, they are messy. Why should systems seek a state of maximum disorder? The answer, though it was a long time coming and quite a struggle to discover, is that the disordered state is by far and away the most probable.

The idea is usually explained by analogy with a deck of cards. There is only one way to deal a perfectly ordered hand in bridge: thirteen cards of the same suit. But there are some 640 billion other, disordered, hands one can be dealt. Disorder is amazingly more probable than order. In a way, this fact can be seen in a positive light: anything is possible—just not very probable. On a molecular level, it is possible that all the molecules of a chair might move in the same direction at the same time and the chair might defy gravity and jump up off the floor—but it is highly, highly unlikely. It is *extremely* likely that the motions of the molecules will remain disordered and any movement of a molecule up will be countered by movement of a molecule down, and the chair will stay put. This tendency of systems to become disordered and stay disordered is strong enough that it explains, along with considerations of energy, the outcome of chemical reactions.

In physical as well as chemical processes, the effect of entropy is to motivate reactions, even those where an input of energy may be required. A classical example is the evaporation of a liquid such as water or alcohol. The indication that the process requires energy is the cool feeling of the skin when water or alcohol evaporates. The water or alcohol absorbs some energy from the body. However, the process occurs quite readily. Using a term that had been adopted by science and fitted with special, narrow meaning, the reaction occurs *spontaneously*. Spontaneity gains a more precise meaning in thermodynamics because, with the machinery of thermodynamics, the tendency of a reaction to occur spontaneously can be calculated and assigned a number. The number it is assigned is called the *free energy*.

What is free energy? It is the energy that results from the difference between the energy gained or lost in a chemical reaction and the amount borrowed or expended because of entropy. Just like a checkbook or a balance sheet, nature tallies how much energy is lost or gained and subtracts the taxes due to entropy. The energy left over is the available capital, or as we say in thermodynamics, the free energy: the energy free to do work. We said that all this reshuffling of energy occurs when chemical bonds are made or broken, but the same considerations come into play when intermolecular attractions are made or broken. Evaporation, the change of a liquid into a gas, is an example of a process that requires energy but can still be spontaneous because so much entropy is gained. Does gas expansion do work? It certainly does in the cylinder of an internal combustion engine.

A liquid might be viewed as a school of fish or a flock of birds: all moving, but in unison, heading first one way and then another, together, with positions in the group seemingly assigned and varying only slightly as the entire group moves. Gas, however, is a swarm of gnats, flitting about individually and in random directions, the group being carried by air current rather than purposeful direction. It is much easier to predict the movement of one fish in a school than it is one gnat in a swarm, and a liquid is more ordered than a gas. Alcohol evaporates from skin, disrupting intermolecular attractions, which requires energy, because it is able to gain entropy by turning into a gas.

Thus a process that absorbs heat can be spontaneous if it creates enough disorder at the same time. On the other hand, if enough energy is expended, reactions that create order can be driven forward. An orderly solid of cake can be created from a disorderly liquid batter with the addition of heat.

In refrigeration, cooling occurs when compressed refrigerant absorbs heat and expands, which disrupts intermolecular attractions. We are all familiar with the sound of refrigerant flowing through the coils at the back of the refrigerator and the compression pump running. But how do they get refrigerators to run off the cigarette lighter outlet in cars? Itty bitty coils and a miniature compression pump? Not quite. But some of the fundamental principles are the same.

## For Example: Refrigeration on the Road

Before we address refrigeration on the road, let's look at the principles behind good old-fashioned household refrigeration. When a fluid is allowed to evaporate on the skin, the skin feels cooler. Try it with rubbing alcohol. Sweating is the body's attempt to cool itself by producing a fluid that will evaporate and cool the skin. Sweat absorbs heat from the skin and uses this heat to break the intermolecular attractions that are holding the materials in the liquid state. A pot of water boiling on your range top and the refrigerant in your refrigerator are experiencing the same effect. The pot of water stays at a steady 212°F (100°C) (assuming you are at sea level) because the heat added from the burner is going into breaking intermolecular attractions between water molecules, not increasing the translational energy. In effect, the pot of water is refrigerating the stove burner by dissipating the heat. Driving this process forward is the need for the ordered liquid to disorder by becoming a vapor.

As an aside, wart removal by liquid nitrogen exploits the same principle. As may be recalled from our discussion of intermolecular forces, the attractions of induced dipoles, dispersion forces, are caused by temporarily uneven distributions in the electron clouds, which produce momentary positive and negative regions. Nitrogen is formed from two nitrogen nuclei bonded together—the electrons have no reason to prefer one nitrogen in the molecule over the other—so dipoles, a permanent separation of positive and negative charge, are not a feature of nitrogen molecules. Cooled and put under pressure, nitrogen liquefies because of the attraction of dispersion forces alone. But for anyone who has felt the cooling of liquid nitrogen, the pain can leave no doubt as to the strength of these forces. The energy absorbed from the skin when liquid nitrogen evaporates causes frost bite severe enough to kill that portion of the skin. If this is the portion of skin that is hosting a wart, then the wart is removed.

The fluid that flows through the coils in the back of a refrigerator removes heat from the interior by the same process. In fact, the refrigerant used in freezers will absorb so much heat that it can freeze skin. The fluid is compressed by the compressor, then allowed to expand through the coils. The compression is driven by an electric motor, and the expansion is driven by entropy: the gas phase is a more disordered state than the liquid phase.

Though the evaporation process is driven by entropy, it still requires energy to happen, so it draws its energy from its surroundings, thus cooling the surroundings in the process. The refrigerant is collected and recompressed into a liquid. The recompressed liquid releases the energy drawn from the refrigerator, which is the heat that can be felt at the back of a refrigerator. After compression, the cycle starts over.

But there is no compressor noise associated with an automobile refrigerator, so what is happening here? Something called the *thermoelectric Peltier effect*, which can also be explained in terms of thermodynamics, as follows.

Some semiconductors allow the passage of electrons more readily than other semiconductors do; therefore, when any two dissimilar semiconductors are joined in a circuit, the electrons experience a change in energy as they go from one semiconductor to the other. When the electrons jump from one semiconductor to a new semiconductor—and then back to the first semiconductor—one of those jumps will be uphill, and the other will be downhill. One of the jumps will require energy, and one of them will release energy. Where do the electrons get this energy? Why not down the road at the junction where it is being given off? That is essentially what happens: heat flows, and one junction gets warm while the other cools. If the warm junction is outside an insulated box and the cool junction is inside, then voilà, you have a refrigerator.

The applications of thermodynamics are many in chemistry, as well as refrigerators, but are singularly useful in understanding phase equilibrium as well as the important concept of chemical equilibrium. Because chemical equilibrium involves phase equilibrium, we will look at phase equilibrium first. The phase before equilibrium.

## Demonstration 14: Zoned-Out Ice

*What Youth deemed crystal, Age finds out was dew.*
—Robert Browning, *Jochanan Hakkadosh*, ca. 1840

We have now discussed the solid phase of matter as well as the gas phase and the solution phase. We have encountered interesting chemistry in each, but this is just the tip of the iceberg, as it were. The next topic for discussion will be the change from one phase to another. Though the physics of phase changes experienced by pure materials—such as melting, freezing, and boiling—has a fascination of its own, the situation becomes even more intriguing when a bit of chemistry is added to the mix. As an illustration of a phase change with a little chemical color, try the following demonstration.

After carefully donning your safety glasses, take up the bottle of freshwater aquarium indicator suggested for purchase in the "Shopping List and Solutions." Add three drops of the indicator to a half cup (120 milliliters) of water in a see-through glass or plastic cup, noting as you do the color of the drops as they exit the container and the color of the drops as they diffuse through the water. The drops will be yellowish-orange coming out of the dispenser and greenish-blue in tap water. The reason for the color change is that the indicator is a mild acid. In water it loses a hydrogen ion to water and in the process changes color. Swirl the solution a bit to mix in the indicator then cover the cup with plastic wrap. Secure the plastic wrap in place with a rubber band or adhesive tape. Wipe the outside of the container clean with a paper towel, and then put the glass in your freezer on a paper towel. Take precautions to keep it away from any food and label it carefully as something not to be consumed. Allow the mixture to freeze solid.

After the mixture is frozen, take it out and look at it closely. You should see that the ice cube formed is clear on the outside and all of the indicator has been pushed together in the middle. The indicator sequestered in the middle will have reverted to a yellow color.

In industry, a technique known as *zone refining* is used to purify metals. In zone refining, the metals are cooled slowly starting on one end of a molten metal rod. The impurities are squeezed to the end of the material where they can be removed. In the ice cube, the water started freezing slowly from the sides of the glass, and as it did, it pushed the indicator out of the mixture and to the center of the ice. Concentrated at the center, the indicator has reverted to the color it was when it was concentrated in its bottle. Zone refining of ice can be seen in regular ice cubes, too. The outside of the ice is clear, but the inside is opaque to solid white. This dense white area in the middle is formed when gases dissolved in the tap water are pushed to the center of the ice cube by freezing.

# CHAPTER 14

## A Whole New Phase

*In the weeks we were together, Hooper became a symbol to me of young England, so that whenever I read some public utterance proclaiming what youth demanded in the future and what the world owed to youth, I would test these general statements by substituting Hooper and seeing if they still seemed as plausible. Thus, in the dark hour before revelry, I sometimes pondered Hooper rallies, Hooper hostels, international Hooper cooperation, and the religion of Hooper. He was the acid test of all these alloys.*

—Evelyn Waugh, *Brideshead Revisited*, ca. 1945

One of the most important concepts in chemistry is that of chemical equilibrium. Unfortunately, it is also one of the more complex and can be confounding. To avoid this dilemma, we will approach the topic incrementally. We have already taken a major step by delving into the subject of thermodynamics. We now take the second step by discussing phase equilibrium, a less complicated and more familiar phenomenon. In fact, we are all so familiar with the different phases of one chemical, $H_2O$, that we have a pet name for each phase—water for the liquid phase, ice for the solid phase, and steam for the gas phase—though the phase we commonly call steam, the stuff you can see, is actually an aerosol of condensed water. True steam is invisible, scalding hot, and rather dangerous. The aerosol you can see is formed when hot invisible steam cools and condenses in midair. What we commonly call steam could more appropriately be called *fog*. Phase change is what happens when ice melts, water boils, and steam condenses to fog.

Phase change is caused by heat: heat flows in for some changes, and heat flows out for others. Heat flowing into a system normally shows up as increased rotational, vibrational, and translational energy, giving rise to temperature increases. However, at a certain temperature (a different temperature for each different substance), so much energy has been added to the system that the molecules are able to break free of their intermolecular entanglements and transpose themselves into the next phase. The rigid structure of a solid eases into the loose state of a liquid, or the loose state of a liquid eases into the free state of a gas. When heat flows out of a system, molecular motion slows, and intermolecular forces become important again. Gas turns to liquid, and liquid turns to solid. However, materials do not have to go through all three stages in order. Depending on the pressure, it is possible for a material to go directly from a solid to a gas or vice versa. An example of a material that goes directly from a solid to a gas at room temperature and under normal atmospheric pressure is dry ice, solid carbon dioxide, $CO_2$. This property of dry ice, and the fact that carbon dioxide is denser than air, leads to its use on the theatrical stage to produce fog effects.

The process of going directly from solid to gas phase is called sublimation. Deposition is the process of going from the gas phase directly to a solid. An example of deposition might be frost: under certain conditions, water vapor in the air will deposit as frost without condensing to a liquid first. Sublimation and its companion process, deposition, are shown in figure 1.14.1 along with the common phases of matter and the processes that connect them.

In figure 1.14.1, the lighter arrows on the inside of the triangle indicate changes for which energy is required, changes termed *endothermic* changes. The heavier arrows on the outside of the triangle indicate changes for which energy is given off, changes termed *exothermic* changes. An endothermic

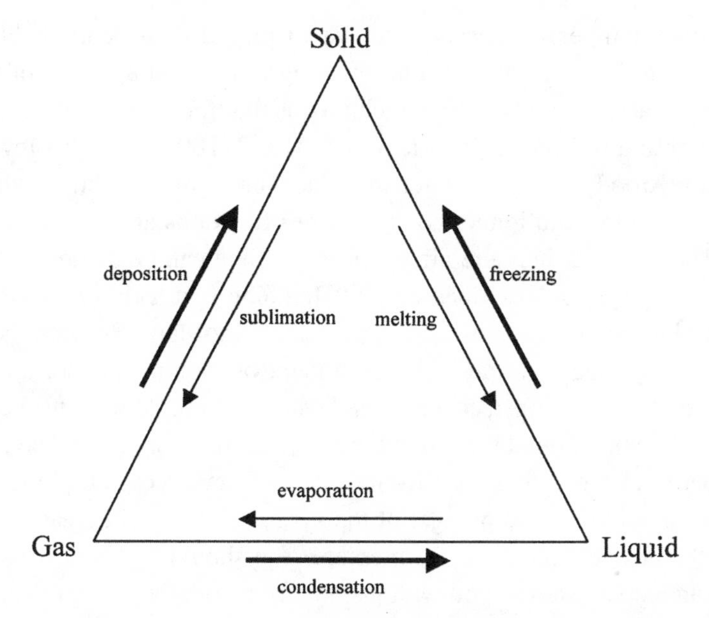


Figure 1.14.1. The normal phases of matter and the physical changes that connect them.

change is one that would take place on a stovetop: melting, evaporation, sublimation. An exothermic change is one that would take place in a freezer: condensation, deposition, and, of course, freezing. The fact that condensation is an exothermic process can be evident in rainstorms: when the gas-phase molecules in the atmosphere condense to rain, heat is given off, which may be sensed as a slight warming of the ambient air.

If that were all there were to the story, this would be a short chapter and chemistry would be considerably less rich. But as we mentioned above, pressure plays an integral role in the wide-ranging behaviors of various materials, which adds another dimension to the discussion. In a mixture, composition also has to be considered, which adds one more factor. Pressure accounts for why ice skating is fun but dry-ice skating is not, and the composition consideration addresses why good pizza goes bad. We'll start with ice skating and have pizza later.

When water boils on the stove, it undergoes a phase change: from liquid to vapor. Bubbles are pockets of vapor that form in the water as it evaporates. The pressure inside the bubbles is the same as the pressure pushing down on the water, otherwise they could not rise and the water would not boil. When pressure is reduced over a liquid, the liquid can boil at a lower temperature because the vapor pressure inside the bubble does not have to be so great. A slight lowering of the boiling point of water is observed at higher altitudes, which means that the maximum temperature of water at a full boil will be less than 212°F (100°C). At this lower temperature, food can take longer to cook thoroughly, which is why some recipes recommend longer cooking times for foods at high elevations.

The way that pressure affects phase change can be demonstrated with the basting syringe described in the "Shopping List and Solutions." Heat a small cloth in hot water, wringing out excess moisture. Remove the needle from the syringe and draw a small amount of rubbing alcohol up into the syringe. With one finger, cover the end where the needle was attached. Wrap the warm cloth around the top of the barrel of the syringe, but leave the end stoppered by your finger uncovered so you can watch the phase change. Draw up gently on the plunger of the syringe, which will create a vacuum over the alcohol and lower the pressure. You should be able to observe the rubbing alcohol start to boil with just the heat from the warm cloth.

We experience phase changes every day with the weather. Even if it doesn't rain or snow, there is often morning dew. The dew point, the temperature at which the moisture from the air will condense on the ground, depends on the ambient pressure. As such, in the past, dew was used as a sort of primitive barometer for weather prediction. The temperature at which water freezes also depends on the pressure. So when we say the freezing point of water is 32°F (0°C), we should really qualify this by saying this is the freezing point under normal atmospheric conditions. But the slight variations we experience in atmospheric pressure are not enough to change the freezing point significantly, so it is justifiable to neglect this qualification in casual conversation. Saying that 32°F (0°C) is the freezing point is also really indicating that this is the point at which water just begins to solidify, so at this temperature there is also some water present, too, which makes ice skating possible. The thin water layer on the ice allows the skates to slide easily. And although ice skating could obviously be accomplished while wearing boots because of the slippery water layer present on the ice at normal temperatures and pressures, ice skating with skates is more fun because it is one of the peculiar properties of water that increasing the pressure lowers the melting point. The design of the skate focuses the weight of the skater on the narrow blade, thereby increasing the pressure for an even more slippery ride. Dry-ice skating would not be fun because, as the name implies, dry ice is dry; that is to say, there is virtually no liquid layer present at normal temperatures and pressures.

Because each substance displays unique behaviors with changing temperature and/or pressure—wax goes from a solid to a liquid when it is heated, egg white goes from a liquid to a solid—it would require columns upon columns of data to describe the temperature/pressure behavior of each substance. It would also require a different table for each substance. In the late 1800s, Josiah Willard Gibbs, a gifted mathematician and physical scientist in the United States invented a device, called a phase diagram, that allows all the data to be displayed on a convenient graph.[1] For example, the phase diagram for $H_2O$ is shown in figure 1.14.2.

The phase diagram displays those regions of temperature and pressure that result in the various phases of water: gas, solid, and liquid. The lines between the phases represent the conditions under which the material is in *equilibrium*; that is, it could just as happily be in either phase, so some amount will be in one phase and some amount will be in the other. The diagram shows that under one atmosphere of pressure, which is our "normal" condition, solid $H_2O$ is in equilibrium with liquid $H_2O$—that is to say, it will begin to freeze or melt—at its normal freezing temperature (32°F, 0°C). If the temperature is higher, if we moved to the right on the phase diagram, we would find that water would be in the liquid phase under normal pressure. If the temperature increases to 100°C while we're still level at one atmosphere, then water would be transitioning into the gas phase.

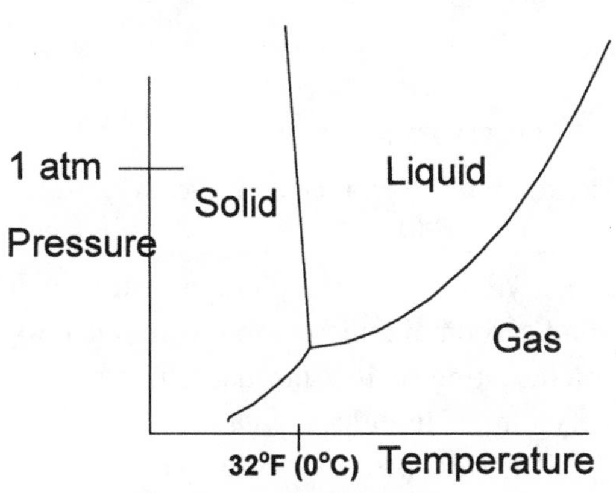

Figure 1.14.2. The phase diagram for $H_2O$.

But not unlike the periodic table, there are other layers of meaning in a phase diagram. If, say, the temperature were 100°C and pressure were higher, then we can tell from the diagram that we would have $H_2O$ in the liquid phase, even though we are at the "normal" boiling point. Also, if the pressure were to go down, the temperature at which the material transitioned from the liquid phase to the gas phase would decrease. This is what happened to the alcohol in our syringe. We can also see from the phase diagram that at lower temperatures water can still be in equilibrium with its vapor, but the vapor pressure will be much lower than at boiling temperatures.

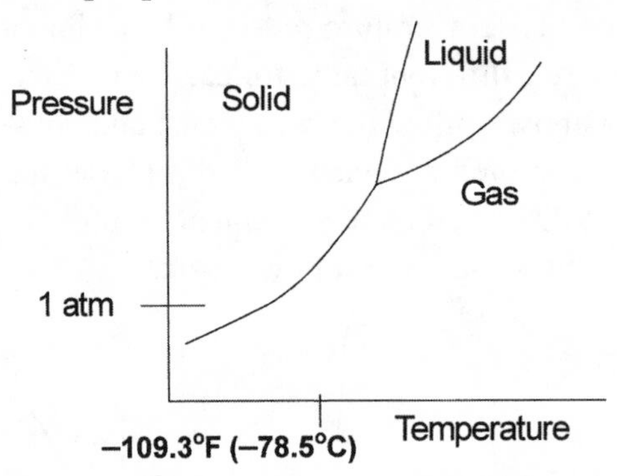


Figure 1.14.3. The phase diagram for carbon dioxide.

The phase diagram of carbon dioxide is shown in figure 1.14.3. From this diagram, we see that at one atmosphere of pressure, carbon dioxide, $CO_2$, is solid only at very low temperatures, and at any temperature above –109.3°F (–78.5°C) it is in the gas phase. So as solid carbon dioxide warms at one atmosphere, it turns directly into a gas (the phase transition called sublimation), which is why solid carbon dioxide is used to simulate fog and why it would be difficult to go skating on it at normal temperatures and pressures: there is no slippery film of liquid, only gas.

It is also possible for a pure material to have several different solid phases. Tin, a material that was once used for organ pipes, can change from a uniformly solid white tin to a powdery gray tin at low temperatures. This phase transition has been termed *tin disease*, or *tin pest.*[2] Ice can solidify into several distinct forms, including ice IX. Reportedly, Kurt Vonnegut, a chemist by training, was unaware of the actual existence of ice IX when he used the name in the plot for his novel *Cat's Cradle*.[3]

The phase changes that occur in mixtures can also be represented by phase diagrams; however, these diagrams tend to be much more complex. A phase is any uniform and stable state of matter, pure or mixture, so each of the different states that a mixture can find itself in qualifies as a separate phase on the phase diagram. There is also the possibility of several phases existing together in equilibrium, and this must be represented, too. Alloys are a class of mixtures that can have distinct phases and serve to illustrate how complicated these phase changes can become. If you have copper, it is possible to have liquid copper or solid copper, depending on the temperature. If you have zinc, it is possible to have liquid zinc or solid zinc, depending on the temperature. But if you have an alloy of copper and zinc, it is possible to have liquid copper and liquid zinc, liquid copper with solid zinc in it, liquid zinc with solid copper in it, or solid copper mixed with solid zinc, each of which is a separate phase. Which situation will be stable depends on temperature, pressure, and the relative amounts of zinc and copper in the mixture—and that's just for two ingredients. If one counts the number of ingredients listed on the label of a bottle of shampoo, it is easy to see how complex a phase diagram can quickly become. In an attempt to simplify the situation, pressure is usually assumed to be constant, at one atmosphere, and the two variables that are tracked are temperature and composition.

Take the simple system of baking soda in water. As baking soda is added to water, the first few pinches go into solution, but as more is added, the mixture becomes saturated, and a solid is present as well as a solution. In other words, solubility depends on composition: if you have a system composed of mostly water and a pinch of soda, the soda goes into solution; if you have a system composed of mostly water but a lot of soda, the soda comes out of solution. If you keep the composition constant, that is, stick with a glass of water that is saturated with baking soda, then you can see how solubility depends on temperature. If you gently heat the solution in a microwave oven, the baking soda will go into solution as the system heats up and the solution will appear clear. If the solution is allowed to cool, the baking soda will come back out of solution.

Another instance in which solubility changes with temperature and composition is in the fine art of distillation. Distillation takes advantage of the fact that the vapor over a solution will generally have a different composition than the solution from which it came. This difference in composition between a liquid and its vapor can be demonstrated with a glass of water that has food coloring in it. If a glass with just an inch or two of water with food coloring is covered with a thin film of plastic wrap held in place by a rubber band and gently warmed on low power for a short time (fifteen to twenty seconds) in a microwave oven, the inside of the glass will be seen to be covered with condensation. The condensation, however, is clear. It is not colored by the food coloring. This color difference is because the composition of the vapor phase differs from the composition of the liquid. The vapor phase is pure water, not the mixture of food coloring and water. The water has a lower boiling point than the food coloring, so it enters the vapor phase more readily, leaving the food coloring behind. There may be some food colorings that are able to vaporize with the water, but most will not and the effect is nice to see. If the experiment doesn't work with your food coloring, get different food coloring (old-fashioned food-coloring drops seem to work best) and try again.

Distillation is used to purify water when necessary and has also been employed for a more nefarious purpose. Ethanol, grain alcohol, can be separated from water by heating up the mixture until the ethanol goes into the vapor phase and the water stays behind. However, not all the water can be removed from ethanol in this manner because of the strong intermolecular forces between water and ethanol. There will always be some water in the ethanol. The best you can get with ordinary methods is 95 percent ethanol (190 proof)—more than adequate for most casual applications.

The general observation that solubility usually increases with temperature can also be illustrated with a sugar-water mixture, but this time with a bit of a twist. While it may be true that a tablespoon of sugar is not soluble in a tablespoon of water at room temperature, gentle heating can cause the sugar to go into solution. This transition, from phase to phase, can again be represented on a phase diagram, such as the one shown in figure 1.14.4.

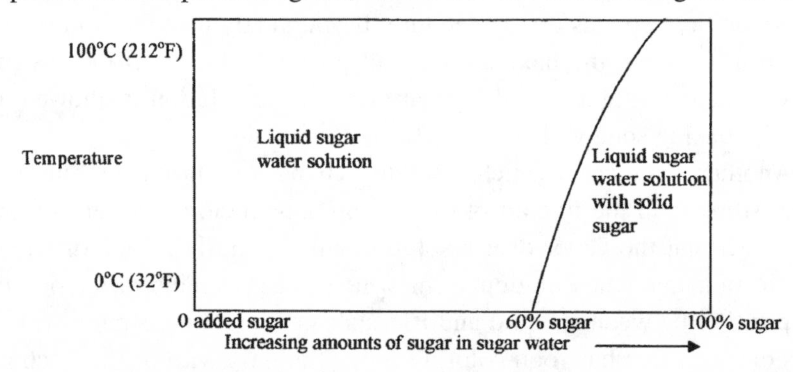


Figure 1.14.4. Phase diagram for the solubility of sugar in water. Note that when we were drawing a phase diagram for a *pure* substance, water or dry ice, the parameters that were varied were temperature and pressure. In this diagram, however, the phase diagram for a *mixture*, temperature and composition are the parameters being varied. Pressure is assumed to be constant.

This diagram tells us that if we have a mixture in which some of the sugar won't go into solution, and the mixture is heated, without changing the composition, the mixture will change phases from a solid-sugar/sugar-water mixture to an all sugar-water solution. In other words, heat up sugar in water and more sugar will go into solution.

This fact may seem a poorly kept secret. The method for forcing more sugar into solution has been known to cooks and bartenders for eons and may hardly seem worth a detailed separate treatment. Hollandaise sauce might also seem a mundane topic for a book on chemistry, but we're going to discuss it here, too, for the following reason: hollandaise sauce is different from sugar water. It displays a behavior that has far-reaching implications—from the London Bridge to Limburger. The difference is in what happens when the hollandaise sauce cools down again.

Hollandaise sauce is a tasty suspension of butter in egg, and when it is warm and freshly prepared, hollandaise pours as a smooth, one-phase mixture. Hot sugar water is a uniform mixture of sugar and water. When sugar water cools down, it remains a uniform mixture of sugar and water. But when hollandaise sauce cools down, something else happens. It separates. Cool hollandaise sauce does not stay together because hot hollandaise is in a *metastable state*, that is, not its lowest energy state but a stable state above the lowest energy state.

A bowling ball balanced atop a stepladder might be an example of a metastable state. The bowling ball is stable, but not at its lowest energy state, which would be on the ground. At the slightest provocation it will seek its lowest energy state and will come tumbling off the ladder. A system that is proceeding to its most stable state so slowly that it is virtually at rest can also be considered metastable. Hot hollandaise sauce is in a metastable state, but the oils will seek to rejoin each other if they can find a route, which is why hollandaise separates.

Metastable states are not confined to sauce. Metastable states can also occur in systems as diverse as medicines and alloys. Many medicines are *suspensions*, a metastable state of solids disbursed in liquid. Metal alloys are mixtures of metals that normally have to be heated to very high temperature in order to mix. If the mixture is cooled slowly enough, separation, such as the zone-refining described above, would probably occur. But if the cooling is done quickly, a metastable solid mixture can result. Of course, a metastable state implies that there is another state, a stable state, toward which the mixture can proceed with time. So metal alloys can change their properties, albeit slowly, with time. In other words, structures composed of alloys, such as bridges, can age. It is not only surface corrosion that must be considered when deciding if a bridge is safe but also the age and nature of the material from which it was made.

But there is no reason for immediate concern on the homeward commute. Not only are bridges carefully monitored for the effects of age, their return to their most stable state from their metastable state can be too slow to observe in several human lifetimes, and most bridges are built to last.

But not so with our next concern: cheese.

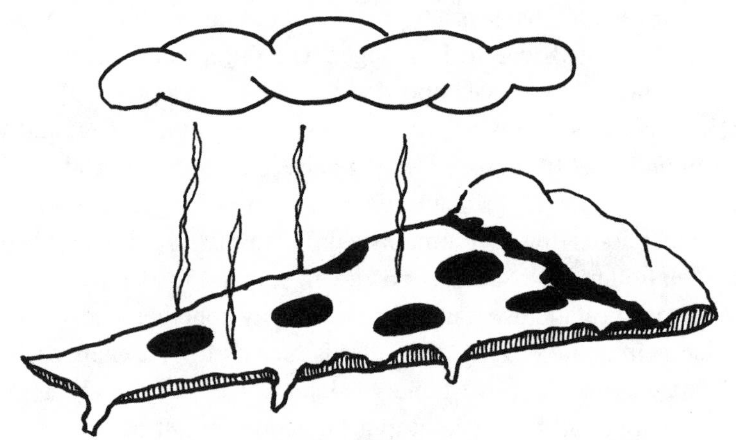

## For Example: When Good Pizza Goes Bad

Many materials we deal with every day are in a metastable state. Medicinal emulsions and suspensions, shampoos, salad dressings, sauces, and condiments like catsup, mayonnaise, and mustard. But of all the metastable mixtures that have been used for foods, perhaps the oldest and most celebrated is cheese.

It is believed that cheese making was discovered by accident when someone tried to store milk in a bag made out of an animal's stomach and as a result, the enzymes in the stomach curdled the milk.[4] It must have been a very brave or very hungry person who tasted that first cheese, but since then, as a species, we have acquired a liking for the stuff, to the tune of about thirty pounds of cheese per person per year in the United States alone. Cheese, a metastable mixture of oil, water, fats, calcium compounds, and milk, is produced when enzymes harvested from the stomach of calves slowly coagulate proteins into solids. The enzymes that start the process of digesting proteins are in all mammalian stomachs, but in the cheese-making process, the enzymes are allowed to carry out their digestion slowly enough that metastable lumps of solid are formed, surrounded by a liquid containing fats and sugars.

The solids are the *curds*. *Whey* is the more liquid part. Cheese can be thought of as a metastable suspension of curds in whey. The texture of a particular cheese is determined by how the curds are chopped and separated from the whey and the amount of time the cheese is cured or aged. When the curds are not separated from the whey and the cheese is not aged, the result is a mushy mixture called unripened cheese, something like cottage cheese. When all the whey possible is squeezed from the curds and the cheese allowed to age, the result is a solid brick called hard cheese, something like parmesan. Aging consists of allowing bacteria time to complete the coagulating process that the enzymes started, which explains why hard cheese keeps well: it has already been attacked by bacteria and has survived. Soft cheeses, like cottage cheese, which have not been aged, have to be kept in the refrigerator.

The action of the bacteria also serves to break down any remaining *lactose*, a sugar found in milk. Cheese has been much maligned as a difficult food to digest, but in fact, hard cheese that has been made using live bacterial cultures should be easier to digest because it does not contain lactose. Adults that are lactose intolerant are often advised to eat cheese as an alternate source of calcium and protein. It has been speculated that cheese's bad reputation for being "binding" may arise from the experience of long-time sufferers of food sensitivities and loose stools who find the normal digestion of cheese "binding" by contrast! There are, however, many cheese products that have milk added back in, so read the label carefully.[5]

Of the thirty pounds per person per year of cheese consumed in the United States, a good deal of it is in the form of melted cheese found atop our new national food: pizza. Pizza has its own yellow-pages heading in most phone directories, a privilege not accorded to apples, artichokes, or spinach. And although the crust and the sauce may be important to good pizza, it must be acknowledged that without the cheese, pizza would just be an open-faced tomato sauce sandwich. The quality and proper properties of pizza cheese is a topic that preoccupies pizza entrepreneurs.

What is the most important property of a good pizza cheese? Taste, of course. But also the ability to melt smoothly, uniformly, and without separating back into curds and whey—which is why pizza is not topped with an unripened cheese like cottage cheese, or a hard cheese like parmesan, but a semisoft cheese: mozzarella.

As we have said, the degree of softness or hardness of a cheese depends on the amount of curds and whey, so we may be able to illustrate the problem with a stylized phase diagram that we could call Little Miss Muffet (LMM): a phase diagram for mixtures of curds and whey.

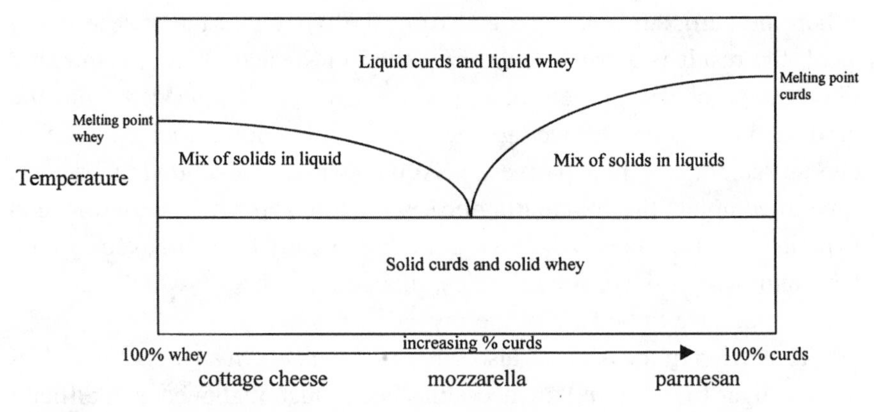


Figure 1.14.5. The LMM hypothetical phase diagram for cheese.

Before interpreting our LMM diagram, we must first acknowledge that LMM is not a real phase diagram. A true phase diagram for a real cheese would be much more complicated. The purpose of this phase diagram is to show how melting and freezing behavior can depend on composition, not to explain the chemistry of cheese. For instance, if a very liquid cheese, such as cottage cheese, is heated in the microwave it will gradually become a uniform mush of curds and whey, and when it is cooled, it will eventually separate into layers of clear liquid whey and white liquid curds. The stuff we call cottage cheese is a metastable suspension of solid curds in liquid curds and whey. A hard cheese might have to be heated to a much higher temperature before it will melt, and it may very well burn before it melts, but if heated gently to a temperature below the melting point, it will start to "sweat" and exude an oily liquid. Because hard cheese, as purchased, is in a carefully crafted metastable state, the cheese will not return to its original state when cooled. After temperature-induced phase separation, the texture of the cheese will have changed when it is cooled; it will have become more brittle or stringy. What is needed for good pizza, however, is a cheese that will not separate like an unripened cheese or sweat like hard cheese. When it came to choosing the best cheese for pizza, people gravitated toward a semisoft cheese, mozzarella, because mozzarella melts smoothly and evenly. We have come to believe that mozzarella is the best-tasting cheese for pizza because that is what we are used to. Actually, another cheese might have done just as well if its phase-change behavior would have been more accommodating.

Another advantage to using mozzarella is that even as a mixture it melts quickly and uniformly at its melting point. For most solid mixtures, the mixture melts over a range of temperatures; that is, the mixture will begin to melt at one temperature, but the entire mixture will not be liquid until it has been heated to some final temperature.

There can be, however, some interim mixture that has just the right composition to show an abrupt, decisive melting point. In chemistry, this single-melting-point mixture is called the *eutectic* mixture. In cheese, it's called mozzarella. Solder is a eutectic mixture.

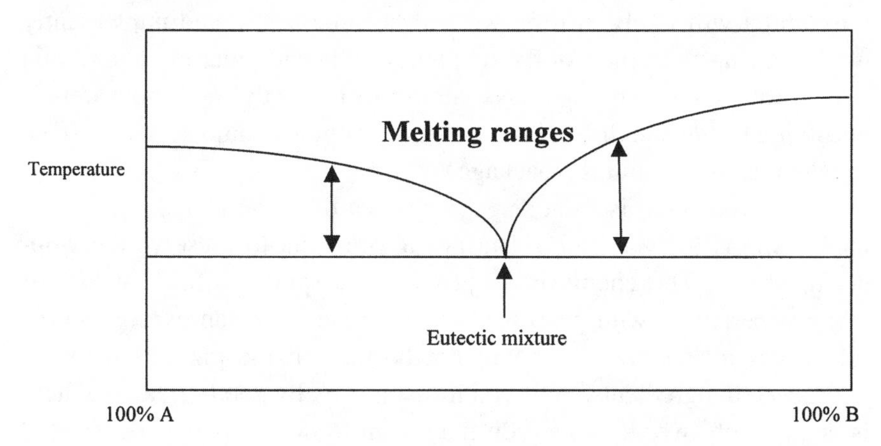


Figure 1.14.6. A generic phase diagram for a mixture showing the melting behavior of a eutectic mixture such as solder.

However, the problem facing the pizza entrepreneurs of America is not just melting cheese but freezing it as well. Pizza lovers of America demand immediate access to their favorite food, which means a form that can be frozen and then quickly prepared in a microwave or convection oven. But cheeses can fare badly when undergoing quick phase changes such as freezing in a freezer and heating in a microwave. Freezing followed by quick heating can result in a crumbly, stringy, tough texture, or burned cheese. The solution has been to use processed cheese, which is a mixture of ground cheese, emulsifying salts, and other stabilizers. Processed cheese allows pizza to be frozen and reheated or even sit out on a table for a fairly extended period of time. The best phase transitions from frozen to melted, however, are achieved by imitation cheese formulated from soybean oil and curds.

To demonstrate the difference in phase change behavior, purchase some shredded real mozzarella and shredded imitation mozzarella topping. Given the existing labeling laws, imitation mozzarella will be clearly marked as imitation, but if there is doubt, a look at the ingredients list should make the difference clear. Place a pinch of each on either side of a paper plate and write "real" and "imitation" on the paper plate near the respective piles. Heat the plate with the cheese samples in a microwave for twenty to thirty seconds. You should see the imitation mozzarella melt smoothly and retain its color and pliability. The real mozzarella will likely turn brown and become brittle and unpleasantly hard. Leaving fresh piles of the real mozzarella and imitation mozzarella out on the counter for two days should result in the real mozzarella's becoming brittle and dry, but the imitation topping should appear as fresh as when it came out of the package.

So if your goal is to defeat the physics of phase change, imitation mozzarella is the way to go. But not if you want to preserve the wonderful aroma. The chemistry is just not the same. Unfortunately, our many experiences with processed cheese have led to an expectation of robustness in our pizza that will not be there if the pizza is made of unprocessed ingredients. So if you invest in a really good pizza, and there is enough left over to store, you may want to do so cautiously. Treated respectfully, true mozzarella may fair just fine; frozen pizza should be thawed before reheating, and it should be reheated on low in the microwave, or perhaps better yet, in a warm oven. But even with these precautions, there may be changes in the texture and taste. So the best idea is to enjoy—eat the pizza when it is fresh and good. If you must, then freeze, but do not be disappointed if that good pizza goes bad.

There remains, however, an additional problem that no amount of stabilizer will fix. And that is that frozen pizza will inevitably dry out in the freezer if it is not properly sealed. For those uninitiated in phase diagrams, the drying out of food in freezers may seem strange. But by looking at the phase diagram for water, it can be seen that at about 32°F (0°C), the temperature of most freezers, solid ice is in equilibrium with a water vapor, though the vapor pressure is much less than one atmosphere. So water vaporizes in a freezer, but because the freezer is sealed most of the time, we might still expect the water loss to be minimal. The difficulty, however, is that frostfree freezers maintain their freedom from frost by circulating air and collecting vaporized water before it has a chance to deposit on the side of the walls as frost. In other words, because water is constantly being removed from circulating air, the water in the freezer is not in equilibrium—it is not in its happy, balanced state with some in the liquid phase and some in the gas phase. More water vapor will be produced from the food as the system tries to restore equilibrium. This drive to establish equilibrium accounts for the fact that ice cubes slowly get smaller in the freezer and poorly sealed frozen pizza can suffer freezer "burn" and will not remain pristine forever. An interesting extension of this manifestation of physical equilibrium is to consider what happens when one hangs one's laundry out to dry in Antarctica. Does it dry? Yes. As long as the air is removing the water vapor from the immediate vicinity of the frozen laundry, the icicles in the underwear will keep evaporating to try to restore equilibrium.

The other kind of equilibrium we now consider, chemical equilibrium, is every bit as common but not always as easily recognized as such and certainly more complex. But complicated or not, it is where we are headed next.

## Demonstration 15: All Things Being Equal

*Without any objective change whatever, variety had taken the place of monotonousness. His host and his host's household . . . began to differentiate themselves as in a chemical process.*
—Thomas Hardy, *Tess of the d'Urbervilles*, ca. 1890

Put on the old safety glasses. Take the copper sulfate solution, prepared as suggested in the "Shopping List and Solutions," and pour about two inches into a clear glass or see-through plastic cup. Take a clear soda bottle, add a half cup (120 milliliters) of water, and then use your funnel to add two tablespoons of baking soda to the water. Shake this solution. While waiting for it to settle, pour two to three inches (5 centimeters) of vinegar into a second clear soda bottle.

Add three or four drops of the baking soda solution to the copper sulfate, making certain no undissolved baking soda is transferred over when you add the drops. You can avoid transferring over undissolved baking soda by using an eyedropper or a straw covered by your finger to pick up drops from the top of the baking soda solution and by being careful not to stir up the sediment on the bottom.

As soon as you add the baking soda solution to the copper sulfate solution, a lightly colored, fluffy precipitate should form and sink to the bottom of the glass. Because you have been careful to add no other solid, this precipitate must be a product of the reaction of the copper sulfate and the baking soda and is, in fact, copper carbonate.

copper ion + carbonate ion → copper carbonate (solid)

If you add a few more drops of baking soda solution, you should see the volume of precipitate increase. The solution above the precipitate should remain blue, which indicates the presence of copper sulfate that has not reacted.

Now add several drops of vinegar to the glass and watch the fizzing that shows the removal of carbonate from the solution. Recall that vinegar is an acid and baking soda is a base. They react to form carbonic acid, which immediately dissociates into water and bubbles of carbon dioxide.

vinegar + baking soda → carbon dioxide + water

If you watch carefully, you will see that the fizzing occurs at the surface of the solution just as the drop of vinegar enters. Keep adding drops of vinegar until the solution stops fizzing. You may need to stir the solution in the glass a little after each addition. As vinegar is added and carbonate ion removed, the precipitate in the bottom should start disappearing.

copper carbonate + vinegar → copper ion + carbon dioxide + water

When the fizzing has stopped and the precipitate has disappeared, add baking soda solution again until precipitate forms again. Then remove the precipitate again with drops of vinegar. This back-and-forth process will continue as long as you keep disturbing the situation by adding carbonate ion or removing carbonate ion by adding vinegar. (Or until the solution becomes too dilute, but the cup would probably overflow before this happened.)

When the baking soda is added initially, copper carbonate comes out of solution until the system is in *equilibrium*; that is, the system is stable and balanced with tolerable amounts of copper carbonate as precipitate and tolerable amounts of copper and carbonate ions in solution. Chemical reactions, however, are not one-way streets; the traffic flows both ways.

So when this equilibrium was upset by removing some of the carbonate ions by allowing them to react with vinegar, the reverse reaction dominated. The system restored its equilibrium by shifting some of the precipitate back into solution. When more carbonate ion was added, the system responded by forming more precipitate again.

Understanding the back-and-forth of chemical equilibrium is of paramount importance to the art of manipulating chemical reactions and is the subject of our next discussion.

# CHAPTER 15

## Equilibrium—Chemistry's Two-Way Street

*Lily sat gazing absently through the blue rings of her cigarette-smoke.*

*"It seems to me," she said at length, "that you spend a good deal of your time in the element you disapprove of."*

*Selden received this thrust without discomposure. "Yes; but (t)he real alchemy consists in being able to turn gold back again into something else; and that's the secret that most of your friends have lost."*

—Edith Wharton, *House of Mirth*, ca. 1905

The preceding demonstration, "All Things Being Equal," was intended to illustrate three significant subtleties concerning chemical reactions. The first is that although we often symbolize chemical reactions with neat little arrows pointing from reactants to products,

$$2\ H_2 + O_2 \rightarrow 2\ H_2O$$

the truth, as suggested several times in previous discussions, is that chemical reactions are really two-way streets. That is, chemical reactions are reversible. If a reactant can become product, then product can become reactant. So although we haven't emphasized this aspect of chemical reactions thus far, whenever a chemical reaction is "done," we are normally left with a mixture of reactants and products, called an *equilibrium mixture.*

Granted, sometimes a reaction will go essentially all to products. When gunpowder explodes, it turns almost entirely into products. "Almost" is an important qualifier here, as gunpowder residue has undone many a miscreant. Other times a reaction mixture may remain virtually unreacted starting material. This problem was faced by Haber in the manufacture of ammonia, as may be recalled from our discussion of gas-phase reactions. Hydrogen and nitrogen gases were all around him, but he had to struggle mightily to bring them together to make ammonia. Many other reactions, however, are like clabbered milk. Clabbered milk is milk in which curds have formed and can be made by pouring vinegar in milk. The curds form immediately when the vinegar is added, but some vinegar and milk remain. The reaction occurs very quickly, but once it is over, the final ratio of curds to uncurdled milk and vinegar stays the same over time. The reaction has reached equilibrium.

The second point we wished to illustrate with "All Things Being Equal" is that saying a reaction has reached equilibrium is *not* the same as saying there are no more reactions taking place. In fact, equilibrium mixtures tend to be very active—with reactants continuously turning into products and products turning back into reactants. This dynamic quality of chemical systems can be illustrated in a straightforward way. Cut a strip from a paper towel and lay it over the edge of a glass containing a little water so that the end of the paper towel just dangles in the water. The water will rise up in the towel only so high and no higher because the water and the towel will eventually establish an equilibrium between the attractions of the towel for the water, the attractions of the water for itself, and gravity pulling the water down. Wait ten minutes to let the towel soak up as much as it will, and then add a drop of food coloring to the water at a spot that is well away from the dangling paper towel. The food dye will disperse by entropy (as we established in our chapter on thermodynamics) and eventually the color will creep up the paper towel, too. The color can creep up the towel because the water molecules in equilibrium are always changing places: some towel molecules are becoming puddle molecules and some puddle molecules are becoming towel molecules. When the dye is added, the new puddle molecules that replace the old ones bring the dye along, too.

But even with all this activity, equilibrium is maintained because reactants change into products and products change into reactants at rates that maintain the same relative amounts of reactant and product. We call this type of equilibrium a *dynamic equilibrium*. Although the net effect may appear constant, it is maintained by steady rates of change. An analogy may be found in a traffic situation. Viewed from a traffic helicopter, the density of cars on a congested freeway may appear fairly constant, but it is the result of constant change: there are new cars entering and exiting all the time.

Which brings us to the third point being illustrated in "All Things Being Equal": as a result of all this activity—products turning to reactants and reactants to products—chemical reactions are flexible. They can respond and adjust to change. If products are removed, the system responds to produce more products until the equilibrium ratio is reestablished. If reactants are added, the system responds to produce more products. In our traffic analogy, this might correspond to the equilibrium between cars on side streets to cars on the freeway:

cars on side streets ↔ cars on freeway

Here cars on the side streets represent our reactants and cars on the freeway represent our products. Our arrow points both ways because cars are getting onto the freeway and off of the freeway. The equilibrium is dynamic and it is reversible.

Early in the morning, there may be a few cars on side streets and a few cars on the freeway, but at 7:30 the number of cars on side streets may suddenly increase.

## cars on side streets ↔ cars on freeway

The system will soon respond, and the number of cars on the freeway will increase until equilibrium is reestablished.

$$\text{cars on side streets} \leftrightarrow \text{cars on freeway}$$

In our above chemical demonstration, we also had a dynamic, reversible situation: copper carbonate precipitate formed, but it also redissolved. When we added vinegar, we removed some of the carbonate from the equilibrium. The solid redissolved in an effort to reestablish the equilibrium of reactant and product.

The ability of chemical systems to respond to stress is captured in *Le Châtelier's principle*, named after Henry Le Châtelier, an influential French chemist of the early 1900s.[1] According to Le Châtelier, whenever a chemical system at equilibrium is stressed, the system will shift in a manner that relieves that stress. This stress can be induced by a change in the quantities of products or reactants, which means we can coax a reaction to produce more of the product we desire. For instance, let's look at a generic reaction:

$$A + B \leftrightarrow C$$

where A and B are reactants and C is a product.

The classic analogy to use at this point is a seesaw, and we will not shy from it because the picture works quite well. When the system is in equilibrium, the seesaw is balanced. This situation is represented by figure 1.15.1a.

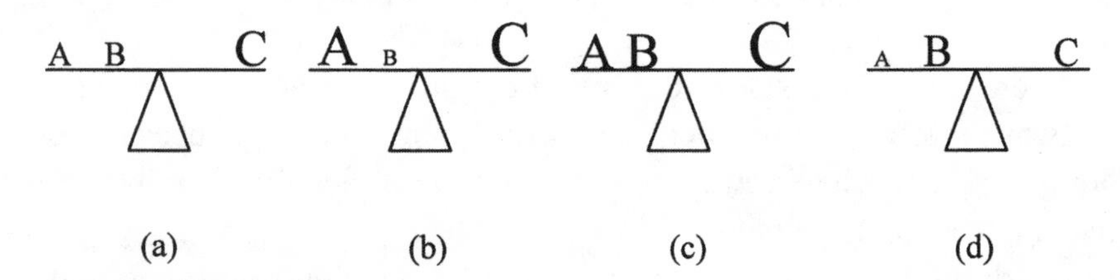


Figure 1.15.1. An equilibrium can be seen as a balanced seesaw. When one side changes, the other adjusts to restore the balance.

If we increase A, the system responds by consuming B to produce more C and reestablish equilibrium. This situation is represented in figure 1.15.1b.

This trick could certainly come in handy if you were in the business of producing C. If, however, your business was to produce B, then there are two tricks you could try. You could increase C to force an increase in B, which would also increase A. This situation is shown in figure 1.15.1c. In our discussion of water softeners, we said that after the water softener is saturated with calcium ions, these ions can be removed by backwashing with large amounts of sodium. This backwashing is an example of increasing the product to shift the equilibrium back toward the reactants.

Alternatively, you could remove some A, which would cause the reaction to shift toward the reactants in an effort to restore equilibrium. As A is replenished, more B is formed and C is depleted. This situation is shown in figure 1.15.1d. All of these techniques have been used at various times in industrial applications. Pretty handy.

The stress on the equilibrium could come from a change in the amount of available energy, too. If a reaction requires energy, energy can be thought of as a reactant. When we dissolved baking soda in water, we noticed the solution became cool. We also saw that warming a baking soda solution increased the solubility of the baking soda.

$$\text{heat} + \text{baking soda} + \text{water} \leftrightarrow \text{baking soda solution}$$

Combining this observation with Le Châtelier's principle gives us a handle on manipulating reactions with heat. If a reaction requires heat, adding heat will cause the reaction to shift to the product side.

$$\text{\Large heat} + \text{baking soda} + \text{water} \leftrightarrow \text{\Large baking soda solution}$$

We can shift the balance the other way by cooling the solution, which, in essence, is removing heat. When we allowed our warm baking soda solution to cool, more baking soda came out of solution.

$$\text{heat} + \text{\Large baking soda + water} \leftrightarrow \text{baking soda solution}$$

So now we have established that chemical reactions are dynamic and reversible, and, as a result, chemical reactions are an equilibrium mixture of reactants and products. We have also seen that sometimes the reaction strongly favors the products (as in reacting gunpowder) and sometimes the reaction strongly favors the reactants (as in Haber's ammonia synthesis from nitrogen and hydrogen). We haven't answered one question, though, and that is, Why? Why are reactants sometimes favored, and why is it sometimes the products? What determines to which side the equilibrium will lie?

One deciding factor is energy. The natural tendency of chemical systems, as with other physical systems, is to move to a state of lower energy. Acids neutralize base. Batteries run down. When all other factors are equal, a reaction will spontaneously adjust to lower its energy and discharge excess energy in the process. But we saw citric acid very spontaneously go into solution, and this mixture *cooled*—it *absorbed* energy—so lower energy can't be the only consideration. To understand the second factor involved, consider the examples we gave at the beginning of this discussion. One reaction went nearly completely to products, gunpowder, and one that stayed nearly completely as reactants—the synthesis of ammonia from nitrogen and hydrogen. The reaction that went to products produced gas-phase products. The reaction that stayed reactants had more reactants in the gas phase. What is the attraction of the gas phase? What does the gas phase have that other phases don't? Entropy. More entropy. What finally determines the position of equilibrium, either favoring the reactants or favoring the products or somewhere in between, is the trade-off between energy and entropy: a situation again best illustrated by analogy. The analogy we choose this time is a trade show.

Trade shows have become so prevalent in our modern world that there are now trade shows for everything and everybody, from computer lovers to campers and from chemists to corrections officers. At a trade show, typically held in a large convention center, people who have items or services to sell that target a trade or special interest group can set up booths to hawk their wares. Because the purpose of the trade show is actually more to advertise than to make sales, the people who staff the booths count their success by the number of people who visit their booths. To attract visitors, the salespeople may offer free goodies for visiting their booth, such as pens, pocket knives, binders, or such. At our equilibrium trade show, we will imagine a very enterprising firm that has decided to offer free vacation giveaways to anyone who visits its booth. Now let's see what the consequences of that action might be.

To begin with, we'll let the booth with the free vacations represent our low-energy state, and we'll set that booth on the products side of the auditorium. We'll now imagine another booth that is charging a five-dollar entrance fee to its booth. We'll call that our high-energy state, and we'll set it up on the reactants side of the auditorium. In between and all around are other booths that are not charging or offering free vacations. Now we'll open the door and allow in the conventioneers. As may be imagined, many will immediately gravitate to the low-energy booth to get their free vacation. This is the situation shown in figure 1.15.2. The reaction being represented here would favor the products because going over to the product side lowers energy in the form of a relaxing, energy-lowering free vacation.

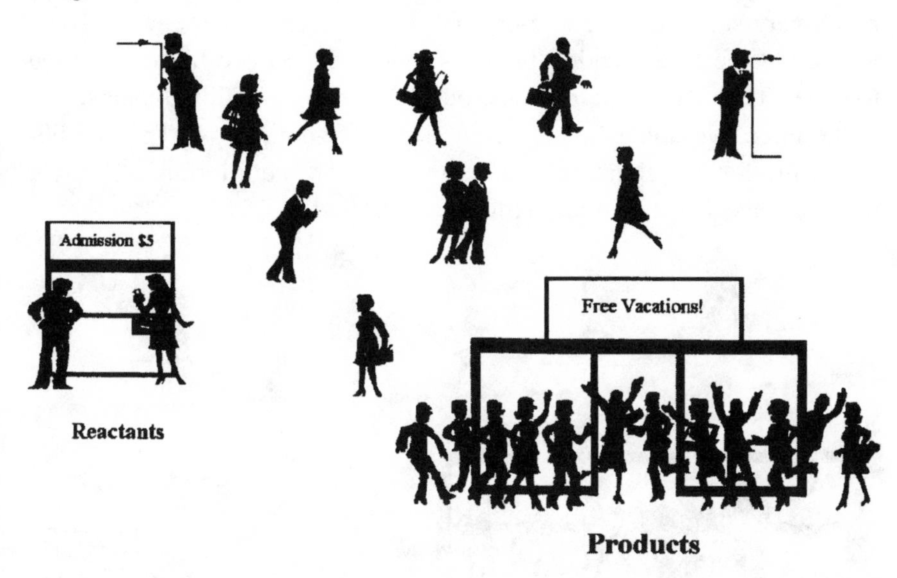


Figure 1.15.2. Attendees at a trade show will tend to gather at the big booth that is offering free vacations because this booth promises a relaxing, low-energy vacation.

However, not everybody will be able to go to the booth with the free vacation because there just isn't enough room around the booth to fit everyone in. Some of the people will be crowded out and will find themselves wandering to the other booths. People will spread over the auditorium because people don't like to be crowded together. We will call this desire to spread out *entropy*.

A molecule of sugar is fairly well located in a handful of sugar. But if that handful is tossed into the ocean, the molecules will spread to the four corners of the globe. This urging by entropy to spread out can cause at least some reactants to take a little heat and go uphill in energy. Because of their desire to spread out, some people will even make their way to the high-energy booth. One or two may even pay the five-dollar charge. The final equilibrium ratio established between the people at the products booth and the people at the reactants booth will be controlled by the desire of the trade show attendees to lower their energy *and* spread out, but in this case will favor the products.

Let's now change the situation a bit. Let's make the free-vacation booth very small and the five-dollar fee booth very large. Fewer people will go to the free-vacation booth because people hate to be squeezed together. They will want to spread out. More people will find themselves at the huge five-dollar booth—even at the five-dollar cost. The equilibrium ratio for this situation will be reversed. The equilibrium will favor the reactants. This situation is illustrated in figure 1.15.3.

Figure 1.15.3. If the free-vacation booth becomes too cramped, people will gravitate to the booth that allows them to spread out, even if the cost is higher.

Chemical reactions can behave in this manner, too. At equilibrium the molecules will distribute themselves between products and reactants in a manner that allows them to minimize their energy and maximize their entropy. But it is always a trade-off. So if there is little to be gained either way for entropy, lower energy will carry the day. But if there is a lot of entropy to be gained, even reactions that require an input of energy can happen spontaneously.

Of course, some reactions have it all—in the explosions in a car engine, gasoline lowers energy by giving off heat and increases its entropy by turning into gases. The heat generated causes the gases to expand, further increasing the entropy.

$$\text{gasoline} + \text{oxygen} \rightarrow \text{carbon dioxide} + \text{water}$$

So the stuff that comes out the tailpipe is basically carbon dioxide and water—with a little soot and NOx mixed in, as previously noted.

So there we have it. The final position for an equilibrium will be determined by the desire to lower energy and maximize entropy: maximum stability. The trade-off is between energy and entropy. In our discussion of thermodynamics, you may recall, we gave this trade-off a name: free energy. Where there is no more free energy to be given off, when there is no more energy free to do work, we have reached a pleasant state of quiescent equilibrium.

And equilibrium always carries the day. Equilibrium may be held at bay for a time, and systems may even enter a metastable state, but ultimately, equilibrium rules. Chemical reactions are able to respond to and adjust to lower energy and maximize entropy because chemical reactions are reversible and dynamic. Being reversible and dynamic means that chemical reactions can be quite pliable and can be coaxed into producing more products or reverting to reactants. Such manipulations are obligatory in chemical industry and the chemical lab, but they are more important than that. Equilibrium concerns us all, every day. As we live and breathe.

## For Example: Breathing Out and Breathing In

In biological systems, equilibrium plays many roles. The one we are perhaps most familiar with is the one we experience some one thousand to three thousand times an hour, twenty-four hours a day, rain or snow, asleep or awake—and that is the equilibrium established by breathing. Free hemoglobin and oxygen are the reactants, and the bound oxygen-hemoglobin complex is the product. We can represent this reaction as

$$Hb + O_2 \leftrightarrow HbO_2$$

In the lungs, the excess of oxygen in the oxygen-rich air forces the formation of more oxygen-hemoglobin complex, $HbO_2$. The blood containing $HbO_2$ then travels to the cells where oxygen has been depleted. If there is relatively little oxygen in a particular cell, oxygen flows into the oxygen-poor cell.

When the atmosphere contains less oxygen, as in the case at high altitudes, there may not be enough oxygen brought into the lungs. When the cells do not get enough oxygen, a person may experience hypoxia, or oxygen deprivation, which can cause headache, fatigue, dizziness, clumsiness, and nausea. The balance has been tipped, and for the body, that is not good. The body, however, operates on the principles of equilibrium, so if there is a deficiency somewhere, the body shifts to make up for the deficit. Obviously, the body cannot make more oxygen, so it does the next-best thing, it makes more of the other reactant: hemoglobin. A person inhabiting the mountains can have half again as many red blood cells as someone habitually at sea level. Of course, making more hemoglobin creates a deficit in the proteins used to make hemoglobin, so this creates pressure on the cells to make more of these proteins, which sets up a whole cascade of equilibrium-restoring reactions, all triggered by the drive to lower energy and maximize entropy.

Can there be a chemical reaction that does not involve a change of energy and entropy? No. If a new material is forming—the definition of a chemical reaction—then there will be making and breaking of bonds, which means there will be an energy and entropy change. In fact, it is difficult to imagine any process, physical or chemical, that involves energy but does not involve entropy. However, there are physical processes that are driven by entropy and involve virtually no energy change. The spontaneous mixing of food dye and water is an example of an entropy-driven process. There is, in fact, another entire class of reactions driven by entropy, and this is what we will explore next.

## DEMONSTRATION 16: ANTIFREEZE AND ANTIBOIL

*Everybody's youth is a dream, a form of chemical madness.*
*How pleasant then to be insane!*
—F. Scott Fitzgerald, *The Diamond as Big as the Ritz*, ca. 1922

In this set of four demonstrations, we will illustrate the four *colligative properties of solutions*. The colligative properties of a solution describe how the presence of solute alters the phase-change behavior of the solvent. The properties are intriguing in that they depend only on how many solute particles are present, not the identity of the particles. An analogy might be made with crowd behavior: crowds tend to behave in ways that depend on the size of the crowd, not the individual natures of the people that make up the crowd. Of course, on some level, the individuality of the people will assert itself, and on some level the nature of our particles is important. But the gross effect, the crowd behavior, is what we will concern ourselves with here.

The colligative properties we will be observing are *freezing-point depression*, *boiling-point elevation*, *vapor-pressure lowering*, and *osmosis*. Put on your safety glasses for all four demonstrations.

To demonstrate freezing-point depression, take two old plastic soda or water bottles that still have their caps and rinse them out. Add a cup (240 milliliters) of water to each and screw the cap on one. Into the other, pour a tablespoon (15 milliliters) of table salt, screw on the cap, and shake the contents. If all of the salt dissolves, pour in a little more and shake the contents again. Keep adding small amounts of salt until no more dissolves and, after shaking, there is some salt remaining on the bottom of the bottle. Identify the saltwater bottle with a permanent marker or a strip of tape, tighten the caps of each bottle, and place them both in your freezer. Check back through the day. You will find that the regular water freezes as you would expect, but the saturated salt solution has such a low freezing point that it will not freeze at home-freezer temperatures, which demonstrates freezing-point depression. You can leave both bottles in the freezer to convince yourself that the salt solution will not freeze.

You may be familiar with the property illustrated in the first demonstration, freezing-point depression, if you have ever used an ice cream maker. In ice cream making, the cream mixture is poured into a container and the container is placed in a bath of salted ice. The salt-ice mixture cools to a point below the freezing point of water, so when the water in the cream comes into contact with the sides of the container, it freezes. A paddle to scrape ice crystals off the sides of the container is turned constantly until the cream and ice form a uniform semisolid. The freezing point of the salt-ice mixture is lowered because of the presence of a dissolved solute (salt) in the solvent (water). Antifreeze works on the same principle. Chosen to be noncorrosive, antifreeze is a material added to radiator water to lower the freezing point so the radiator water will not readily freeze.

A second colligative effect, called *boiling-point elevation*, can be demonstrated with vinegar. Vinegar displays colligative effects because it is a solution of acetic acid in water. You can measure the temperature of boiling vinegar with the large-range thermometer suggested for purchase in the "Shopping List and Solutions." These thermometers are not always that accurate, so to see the difference you may have to compare the temperature of boiling vinegar to the temperature of boiling water by putting the thermometer first in one and then in the other. You should find that vinegar boils at around 216°F (102°C) instead of 212°F (100°C), the temperature at which pure water boils. While your thermometer may not read this exactly, the vinegar should still show a higher boiling point than the water does. The boiling point of the solution is higher with the solute than without the solute, which is an example of boiling-point elevation.

Our third demonstration involves the decreased ability for a solution to evaporate when compared to the plain solvent. Once again, this property, called *vapor-pressure lowering*, does not depend on the chemical nature of the solute but simply on how much of the solute is dissolved in the solvent. Take two spoons and crush three aspirin tablets. Take two small, clear cups and add the crushed aspirin to one of the cups. Pour a half cup (120 milliliters) of rubbing alcohol into each. It is necessary to have the solution levels the same. Swirl the aspirin-alcohol cup to mix the contents. (Some may remain undissolved.) Label the cups to show which one contains the aspirin and then set both cups in a similar environment so that they both get the same amount of heat, air currents, and so forth. Side by side on a counter works well. In about an hour, you should observe an obvious difference in the levels in the cups. Although solvent is evaporating from both cups, it is evaporating more slowly from the aspirin cup. The same effect can be observed for vitamin pills or iron tablets dissolved in alcohol. Thus, it is the presence of these solutes that causes the effect, not the nature of the solutes.

To show that the number of solute particles makes the difference, try the above aspirin demonstration again, but this time set up three or four cups and vary the number of aspirin in each cup. After allowing them to evaporate for about an hour, you should end up with three or four distinctly different solvent levels, ranging from a low with the cup containing the fewest aspirin to a high containing the cup with the greatest number of aspirin.

The last colligative property to demonstrate is *osmosis*. The process of osmosis is rather specialized in that it requires a *semipermeable membrane*: a material that will allow solvent molecules to pass through but not solute molecules. In osmosis, solvent flows from one solution to the other through the semipermeable membrane. If both containers are open to the same atmospheric pressure, the direction of the flow is from the more dilute solution and into the more concentrated solution.

To demonstrate solvent flowing into a concentrated solution, drop a raisin or other piece of dried fruit into water and check it after about an hour. The raisin will have swollen up as the pure solvent, water, traveled through the semipermeable membrane—the raisin's skin—and into the raisin's cells in an effort to dilute the concentrated contents of the cells. To demonstrate the flow of solvent from the less salty solution of the cells into a salty brine, cut two slices of raw potato as thinly as you can. Even a very thin slice of raw potato will feel stiff and crisp because the cells that make up the potato are completely filled with water. Put one slice in a strong saltwater solution made from one tablespoon (15 milliliters) of salt in one half cup (120 milliliters) of water. Put the other slice in pure water. Wait half an hour and then pull out the potato slices and examine them. The one that was soaking in pure water will still be crisp. The slice that was soaking in salt water will be limp. Compare it to a freshly cut slice of potato. It is limp because some of the water left the cells and went into the more salty brine.

The seemingly effortless movement of the solvent in osmosis is captured by the expression "learning by osmosis." Learning by osmosis implies that the knowledge can be absorbed by proximity alone and does not require effort by the learner. But learning is never truly effortless. Whether in learning or chemistry, there are forces behind osmosis, as well as the other colligative properties, as we will discover in the pages that follow.

# CHAPTER 16

## Colligative Properties—Strength in Numbers

*In the experimental sciences what we know we call the laws of inevitability, what is unknown to us we call vital force. Vital force is only an expression for the unknown remainder over and above what we know of the essence of life.*

—Leo Tolstoy, *War and Peace*, 1866

When dealing with chemical reactions, the first questions usually have to do with the identity of the material. What's the reactant? What's the product? What's that smell? When it comes to the *colligative* properties of solutions, however, the question is how much, not what kind, of solute is in the solution. In other words, it may matter to you whether you put sugar or brandy in your tea, but as long as it is the same number of molecules from sugar as from brandy, it would not matter a whit to the colligative properties of the tea.

The colligative properties of solutions are physical properties, not chemical properties, because the chemical nature of the solute and the solvent remain unchanged. In truth, in solutions where there are significant intermolecular forces between particles, different solutes may behave a bit differently, but the general effect will be the same. These effects are germane to many chemical processes and control circumstances as diverse as the functioning of the living cell to the preservation of life on the highway.

The colligative properties of solutions are phase-change properties and include an odd collection of effects. Freezing-point depression refers to the observation that solutions have a lower freezing point than the pure solvent. The more particles of solute added, the lower the freezing point. Road crews make use of this effect in the winter when they spread salt on the roads to dissolve the ice. As we mentioned in our discussion of phase changes, a sample of ice will be coated with a thin film of water because of the equilibrium among the phases. Salt sprinkled on ice will dissolve in this water film and make a very concentrated solution. Equilibrium, as we have pointed out, is dynamic, meaning that the molecules in the solid phase are constantly trading places with the molecules in the liquid phase. In this instance, however, the situation is a bit different. When the solid molecules venture into the solution, they create a solution that is no longer at its freezing point, so they do not turn back into a solid. The balance has been tipped toward the liquid phase, and gradually more and more molecules find themselves in solution. Sand is also sometimes used on slippery roads, but sand is apt to be used in situations involving snow or slush, where traction is more of the problem. When the goal is to melt ice, the salt trucks are called out.

Boiling-point elevation refers to the fact that solutions boil at a higher temperature than the pure solvent. The more solute, the higher the boiling point. A solute, such as antifreeze, added to a solvent, such as radiator water, will lower the freezing point of the solution in the radiator so that it won't freeze quite so readily. It turns out the antifreeze could also be called *antiboil* because the same substance that helps to keep radiators from freezing helps to keep radiators from boiling over, too. Two cautions need to be added here. First, *antifreeze* and *antiboil* do not imply radiators won't ever freeze or won't ever boil, just that a lower or higher temperature must be attained to achieve freezing or boiling. Second, the antifreeze one adds to a radiator is a poisonous brew, so it is *not* a substance with which to experiment.

When we talk about vapor-pressure lowering, we are talking about the behavior of fumes. Any volatile solvent is going to have fumes associated with it, such as the pungent fumes of gasoline or alcohol. In the parlance of chemistry, these fumes are referred to as the vapor above the solvent. The amount of vapor associated with a solvent will depend on the temperature of the solvent, which is why many people associate gasoline fumes with hot summer days more than with chilly winter mornings. The vapor pressure also depends on the amount of solute that is dissolved in the solvent.

Osmosis is the tendency for solvent to flow into salty solutions to dilute them. Osmosis is responsible for the revival of wilted celery when soaked in pure water: water flows into the celery to dilute the salty cells. Osmosis is responsible for pickles pickling: water flows out of the pickles in an attempt to dilute the salty brine. How does the taste get into the pickle? Osmosis is also striving for equilibrium, and equilibrium situations in chemistry are dynamic. At equilibrium, the flavorful molecules will be redistributed between pickle and brine. Recall the demonstration with the paper towel and the food dye: the paper towel was allowed to take on its equilibrium load of water from a puddle, and then food dye was added to the puddle. Because equilibrium is dynamic, some food dye eventually found its way into the towel. At equilibrium there will be more water outside the pickle cell than inside, but the flavoring in the brine will have found its way into the pickle.

Cells exist, and enable life to exist, because the cell membrane is semipermeable: it allows in what the cell needs, allows out what the cell doesn't need, and keeps the contents of the cell from spilling out or drizzling away. An example of the action of a cell's semipermeable membrane can be found by anyone who takes the time to soak in a bath: skin puckering. When asked, most people will say that bathtub skin puckering is caused by water flowing out of the skin. In truth, the folding and bunching of the skin is caused by the skin swelling up with the water it has absorbed. The fluid in the cells is a very concentrated concoction of solutes: proteins, salts, sugars, and other specialized molecules, so the water in the bathtub flows into the cell in an attempt to dilute the cell's contents.

But bathtub amusements are not the only use the body has for osmosis. The body uses the presence or absence of certain solutes to stimulate or inhibit various processes that are necessary for cell growth, reproduction, and repair. Around the turn of the nineteenth century, the German scientist Jacques Loeb managed to stimulate unfertilized sea urchin eggs to develop to the larval stage by changing the salt concentration of their surrounding solution, thereby establishing life as a physico-chemical process rather than the result of a mysterious vital force.[1]

Nonetheless, colligative properties of solutions still remain a bit mysterious in and of themselves. The current consensus seems to be that there are two major factors driving the observed colligative behaviors. One is that the physical presence of the solute in solution interferes with the normal function of the solvent particles. For instance, the vapor pressure is lower because solute particles take up some of the surface sites from which the solvent would depart into the gas phase. This inhibition of evaporation causes vapor-pressure lowering—and thus boiling-point elevation—because boiling occurs when the vapor pressure of the solvent is equal to the ambient pressure and bubbles of vapor can escape. Solute molecules can also interfere with the intermolecular attractions that facilitate freezing and thereby lower the freezing point. But with these effects accounted for, the major driver for the observed colligative properties is entropy.

As you may recall, systems tend toward a state of maximum disorder, that is, maximum entropy, and a dilute solution is more disordered than a concentrated solution. Now consider two extremes: one in which there are only two particles—one solute particle and one solvent particle—and one in which there is one solute particle in a sea of solvent. The first situation, the one with one particle of solute and one particle of solvent, is considered very ordered because it would be easy to locate the solute particle. The other situation, the one with one solute molecule in a sea of solvent, is considered very disordered because it would be impossible to locate the solute particle. The more dilute a solution is, the more it is disordered.

As we affirmed in our discussion of thermodynamics, systems naturally tend toward a state of maximum entropy. Physical and chemical systems alike favor disorder. It is their natural state. Gases expand. Dyes distribute. Mixtures do whatever they can to dilute. Solutions do not want to lose solvent and become concentrated because there is more entropy in being dilute. The vapor pressure of a solution is lowered so the solution will not lose solvent and become more concentrated. The boiling point is elevated for the same reason. There is also more entropy for a liquid solution than for a solid, so solution freezing is anti-entropic. Hence, solutions experience freezing-point depression. In osmosis, the motivation is to seek a uniform, dilute solution as a means of increasing entropy.

From pickles to celery to cells: if there is a vital force, it is entropy.

## For Example: Kidneys and Chemistry

What do kidneys have to do with pickles and celery? As it happens, quite a bit. Colligative properties of solutions figure into body chemistry in a big way because the body is just one big salt solution partitioned into cells by semipermeable membranes. The semipermeable membranes of the cells, however, have a much bigger job to do than letting water into limp celery or pulling water out of pickles. Blood bathes the cells in a solution of many particles beside water, and the cells themselves contain a soup of many particles in water. Some of these particles are relatively large, some are relatively small. Some are ions, some are neutral. Some are polar, some are nonpolar. Some are supposed to be there, some are not. To regulate the composition of cell contents, the blood must be rich in what the cells need and deficient in whatever the cells consider waste. The tricky part is deciding what is waste and what is not: what stays and what goes.

For instance, urea, the product of protein digestion from which the term *urine* is derived, must be removed. Other waste products of metabolism must be removed. Any toxins produced by bacteria must be removed. Any drug residue or other unusable material must be removed. Any excess hormone must be removed. Glucose, on the other hand, should not be eliminated and proteins should not be secreted. Vitamins need to be saved, as does calcium and a certain amount of sodium and other minerals. However, water must be regulated. Too much water in the blood would be bad because if blood were too dilute, then not enough nutrients would be transported to the cell. If blood had too little water, then the physical process of pumping the blood around would be impaired. The job of the kidneys is to sort among the various materials that come floating by and choose what to keep and what to flush.

One straightforward mechanism employed by the kidneys is good old-fashioned filtration. Water and small molecules are squeezed out of the blood by applying pressure, as in an espresso machine. The solution that comes through the filter (the coffee, as opposed to the grounds, in our espresso example) is called the *filtrate*. Red blood cells are too large to be filtered at this point and should stay out of the filtrate. Any red blood cells visible in urine indicate a possible leak in the lines and should be brought to the attention of a physician.

However, a filter isn't very discriminate, and some of the materials small enough to pass through the filter actually need to be retained. The kidneys compensate by having special proteins called *transporters* on the kidney membranes that recognize and reabsorb molecules such as certain sugars and proteins. Of course, the body has developed transporters only for those substances that it has manufactured itself or routinely takes in from the environment. Human-developed drugs are categorized by the body as waste and, as such, are eliminated. Therefore, drug dosage has to be timed and regulated to keep adequate quantities in the blood.

The blood must not be allowed to become too acidic or too basic—the pH of the blood must not be allowed to go too high or too low—so the carbonate and bicarbonate ions must be carefully regulated. If there is too much carbonate, it will be skimmed off and removed, and if there is too little, there is a mechanism for restoring this equilibrium, too.

A product of kidney cell metabolism, like the metabolism of all cells, is carbon dioxide. As we have noted before, carbon dioxide reacts with water to form carbonic acid, some of which becomes carbonate in the blood. Under other circumstances, carbon dioxide would be transported to the lungs and exhaled. But if it is needed by the blood to restore pH, this carbon dioxide can be retained.

Of all these intricate and specialized mechanisms for partitioning materials between the blood and the waste stream, osmosis is extremely important. However, if osmosis were allowed to proceed without regulation, our cells might swell or collapse like ungoverned celery and pickles. Luckily, there are hormones produced by the body that affect the permeability of the membranes across which osmosis takes place. One such hormone is the antidiuretic hormone, ADH.

ADH is secreted by the pituitary gland, a gland located at the base of the brain. The pituitary gland is also responsible for secreting other hormones that regulate growth and metabolism. When the pituitary gland detects high levels of sodium in the blood, ADH is released so that water will be retained by the kidneys, the blood will be more dilute, and there will be less water loss by the cells—a mechanism that no doubt worked quite well when we were running across the savannah and eating the diet that we had evolved to eat. Unfortunately, we no longer run as much as we used to, and we tend to have much more sodium in our diets than would have ever been available to us on the savannah. Therefore, water retention can be abnormally high, which can upset many body systems. Certainly weight-loss regimens can be defeated by either eating diet meals made tasty with too much sodium or by refraining from drinking water in the belief that excess water will be retained. In truth, extra water dilutes the blood and convinces the body that it is okay to shed water rather than retain it. In addition, dehydration brings about fatigue (when the body senses a lack of water, it attempts to conserve what it has), which makes exercise less likely. So while drinking extra water on a diet may seem counterintuitive, it could help increase the metabolic rate.

Metabolic rates are not the only rates that can be manipulated. The rates of all chemical reactions, from the most explosive to the most sluggish, are described by a set of basic principles, and it is these that we will look at next.

## DEMONSTRATION 17: KICK IT UP A NOTCH

*I saw a telegram handed to a two hundred pound man at a desk. . . . He gave the boy a nickel, tore the envelope and read. Then he yelled "Good God," jumped for his hat and raincoat, ran for the elevator and took a taxi to a railroad depot. . . . (I)t was like a set of crystals in a chemist's tube and a whispering pinch of salt. . . . I know a shoemaker who works in a cellar slamming half-soles onto shoes, and when I told him, he said: "I pay my bills, I love my wife, and I am not afraid of anybody."*

—Carl Sandburg, *Smoke and Steel*, ca. 1922

*Syme sprang to his feet, stepping backwards a little, like a chemical lecturer from a successful explosion.*

—G. K. Chesterton, *The Man Who Was Thursday*, ca. 1908

Two major factors that affect the rate of a chemical reaction—the *kinetics* of the reaction—are concentration and temperature. The effects of both are easily demonstrated using items from our "Shopping List and Solutions."

For the first demonstration, put on your safety goggles and pour a half cup (120 milliliters) of household bleach straight into each of two small clear cups. Be careful that the bleach doesn't splatter, just as when doing the laundry. You may want to wear gloves to protect your hands. Place both cups on a sheet of white paper so the color is easy to observe. Make sure you have a watch or clock in a convenient viewing location.

Add two drops of phenol red indicator (found in the swimming pool test kit) to the first cup. Immediately after the addition to the first cup, add four drops of methyl red indicator to the second cup. Swirl both cups gently to ensure mixing. Set both cups down on the white paper and note on the paper the time at which the drops were added. The red color present in each cup will fade slowly over the next five to ten minutes as the bleach attacks the dye molecules that are coloring the solutions. When the red color is gone from the cup with two drops, note the time on the paper and turn your attention to the cup with four drops. Note the completion time for that reaction on the sheet of paper. You can empty everything down the toilet bowl, but make sure to avoid splattering the bleach.

You can watch the color fade because chemical reactions take time. Some reactions are extremely fast and some are extremely slow, but they all take time. The bleach reaction is paced so that we can observe the process without growing old ourselves. You should have observed that the cup with four drops of dye takes longer to react than the cup with two drops. However, it is not twice as long, even though there is twice as much dye to attack. In general, the rate of reaction is affected by varying the concentration, but not always in a direct way.

The effect of temperature on reaction rates can be demonstrated with two tablets of effervescent antacid, two cups, and tap water. Into one cup, place a half cup (120 milliliters) of cool tap water from the faucet. In the other cup, place an equal amount of hot tap water from the faucet. Drop one tablet into each cup at the same time. The fizzing action is clearly more vigorous in the hot water than in the cool water. In this case, the higher temperature helps in two ways. It forces more bubbles out of solution (which is the same reason you're cautious about opening a warm can of soda), and it increases the reaction rate because molecules at a higher temperature move around faster, find each other more often, and hit each other harder when they do. This effect and other principles concerning chemical kinetics is the subject of the following discussion.

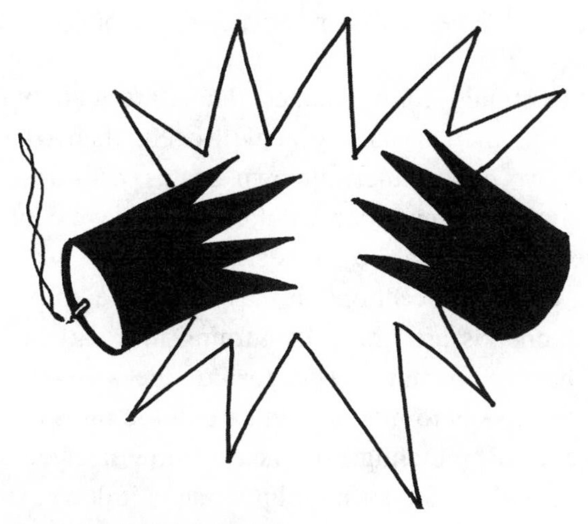

# CHAPTER 17

## Chemical Kinetics—A Veritable Explosion

*Roger, you blessed innocent! . . . Life will no longer be tranquil with a girl of nineteen round the place. You may fool yourself, but you can't fool me. A girl of nineteen doesn't REACT toward things. She explodes. Things don't "react" anywhere but in Boston and in chemical laboratories. I suppose you know you're taking a human bombshell into the arsenal?*

—Christopher Morley, *The Haunted Bookshop*, ca. 1940

Thermodynamics, as noted earlier, deals with the effects of heat and entropy on equilibrium and is useful for predicting the extent to which a reaction will occur. But like an obstreperous child avoiding chores, thermodynamics may tell you the reaction *will* take place, but not *when*. For example, thermodynamics predicts that graphite is a more stable form of pure carbon than diamond, so all diamonds will eventually turn to graphite. But before you start adjusting your investment portfolio, be advised that this transition may take billions of years.

Thermodynamics should not be blamed. There are many pursuits in which it is easier to predict that an event will occur than when it will occur. For instance, we can predict our own deaths with absolute certainly, but imagine how differently we would behave if we could predict when! But regardless of predictability, it is critical to know reaction rates when manipulating chemical reactions. It may not come up in everyday conversation that diamonds are gradually turning into carbon dust, but explosions make the nightly news. The goal of the study of reaction rates—chemical kinetics—is to measure and study the rates of chemical reactions with an eye to predicting how fast or slow a given chemical reaction will proceed and how to speed it up, or slow it down, as needed.

The word *kinetics* implies movement—as in kinetic art or kinesiology, the study of human muscles and motion. The word *kinetics* was chosen to describe the study of reaction rates because reaction rates rely on the movement of molecules. Scientists have been measuring reaction rates since at least the mid-1800s in Europe, but the explanations of the observed rates had to wait for the general acceptance of a theory of molecular motion. The kinetic theory of molecular motion states that molecules and atoms are always in motion, though the motion may be as subtle as the vibrations of ions in a crystal lattice or as frenzied as gas-phase molecules clocking around 1,640 feet (500 meters) in one second.

At normal atmospheric pressures, however, the distance traveled by a gas molecule in one second would not be in a straight line: the density of atmospheric molecules under normal terrestrial conditions is such that one molecule in the gas phase will experience some seven billion collisions in each second. All these collisions are good news for chemists who study reaction rates because the rate of collision is related to the rate of reaction: before anything else can happen in a reaction, molecules must first come together. They must collide.

The idea that reaction rates are tied to the rate of collision is aptly called the *collision theory of reaction rates*. But nature's mysteries are not so easily solved. During a chemical reaction, the electron orbitals (that is, the electron clouds) on individual reactants overlap and coalesce, the way two bubbles come together and merge into one. Once the conditions are right, it takes about a quadrillionth of a second for the electrons in the individual orbitals to readjust themselves into the orbitals around the products. But if all reactions took place as fast as collisions, then food would cook before you could blink; paint would dry before it was spread; and we would have died as soon as we were born because the metabolic reactions in our body would have burned us up. So why don't all reactions happen in fractions of seconds? The answer is in the phrase "once the conditions are right."

To begin with, gas-phase reactants may encounter each other billions of times per second, but solution-phase reactants can be held apart by a sea of solvent and encounter each other far less often. Dropping some food dye in water and dropping some food dye in glycerin should serve to convince you that the rate of diffusion—the movement of solute in a solution—can vary significantly from solvent to solvent. (For a better show, view these diffusions from the side.) But then again, viscosity does not always inhibit a reaction. Sometimes viscosity can enhance the reaction rate by holding the reactants together.

But even when the reactants are able to collide, not all collisions result in a reaction, just as not all dates end in marriage. Once again, the conditions have to be right. For one, the orbitals have to be aligned just so and the collision has to occur with sufficient energy. A bumper-to-bumper, two-car encounter at five miles per hour may not even cause a dented fender. But a side-on collision at forty miles per hour could crush in the side of a car. So in order for a reaction to occur, the reactants have to come together, and they have to have a specific orientation. They have to have the right energy, too.

Given what we know about molecules and their motion, we can do certain things to manipulate reaction rates. For one, we can control concentration. The idea is straightforward—the bigger the crowd, the greater the chance for reaction. But as noted in the directions for the first demonstration given above, the increase in rate is not always directly proportional to the increase in concentration. If the reaction, for instance, is not just one step but a series of steps (as most reactions are), then there can be a complex response to concentration increases. For some reactions, higher concentrations of some reactants can actually impede a reaction.

Another factor that can influence the rate of a reaction is the physical state of the reactants. Ground-up chalk will dissolve much more quickly in vinegar than a solid stick of chalk. The phase of the reactants can also influence the rate of the reaction: liquid gasoline burns; gasoline vapor explodes. Foreign surfaces are always present in a reaction, even if it is just the surface of the container, so they have to be considered, too. Sometimes these surfaces increase the rate of reaction by holding one reactant in an advantageous orientation. Sometimes surfaces impede the reaction by promoting a reverse reaction or by cooling the reactants on collision.

Temperature is virtually always a major factor because it is important on so many levels. The vast majority of reactions speed up as the temperature increases because raising the temperature increases the speed at which the molecules are zooming around, which increases the number of collisions and the energy of the collisions. One of the mammalian responses to infection or injury is to raise the body's temperature so that (among other effects) the immune system reactions and repair reactions will occur more quickly. Increasing the temperature usually increases a reaction rate, but not always. If a certain step in a reaction produces heat, then increasing the temperature, adding *more* heat, can drive this reaction step backwards and end up slowing the reaction down. Predicting and manipulating reaction rates is rarely straightforward.

The state that is halfway between reactants and products, called the *activated complex*, must also be considered. The intermediate might be thought of as the batter when making a cake. You start with shortening, eggs, flour, sugar, and milk (the reactants), but before they become the cake (the product) they have to go through an intermediate state, the batter.

Every reaction goes through an intermediate, however short lived. The reactants have to come together and form a complex in which the orbitals are rearranged and the product formed. It always requires energy to form this activated complex, this intermediate, though the amount of energy may be very small or very large. Chemists speak of an *activation barrier*, meaning the minimum amount of energy that is required to form the intermediate and then kick it down the hill to the product side. This activation barrier can be understood in terms of any human endeavor: it takes energy to go from idea to finished product, no matter how desirable the finished product may be. In chemical reactions, it requires energy to get over the initial hump in going from reactants to products, even if energy is given off by the overall reaction.

At other times, however, the situation can be helped (finally some good news); sometimes a catalyst can be found that will speed up the desired reaction. A catalyst is a substance that lowers the activation energy required for the reaction. A catalyst does not change the amount of product produced; it just makes the reaction faster. A catalyst would be like a moving sidewalk at an airport: it doesn't bring the terminals any closer together, but it does reduce the amount of energy required to move from terminal to terminal and it speeds up the rate at which you can walk there. To further extend our analogy, the moving sidewalk conveys travelers from terminal to terminal but doesn't leave the airport itself. A true catalyst enters into the reaction, but it is not used up in the reaction.

Enzymes are catalysts. The dietary supplements that help the body digest dairy products or roughage such as beans are, in fact, enzymes. Enzymes are sometimes added to detergent to speed the breakdown of stains. A catalytic converter in a car speeds the breakdown of nitrogen-oxygen and carbon-hydrogen compounds into more innocuous nitrogen gas, oxygen gas, water, and carbon dioxide. Catalytic converters have to make these reactions very fast to keep up with the exhaust of a standard automobile.

For some catalysts, the mechanism for the catalytic action is fairly well understood. Biological enzyme catalysts, for instance, may be structured so that intermolecular forces will hold reactants together in an optimum position for reaction. These enzymes act like hands screwing a lid on a jar. To screw a lid on a jar, a jar and the lid could be tossed together in a bag until they eventually found each other, and the lid landed just right and twisted. But if this were the only way a jar could gain a lid, it would take a very long time to screw lids on jars. Two hands can bring the jar and the lid together and get the job done a lot faster. A sketch of an enzyme acting in this manner is shown in figure 1.17.1.

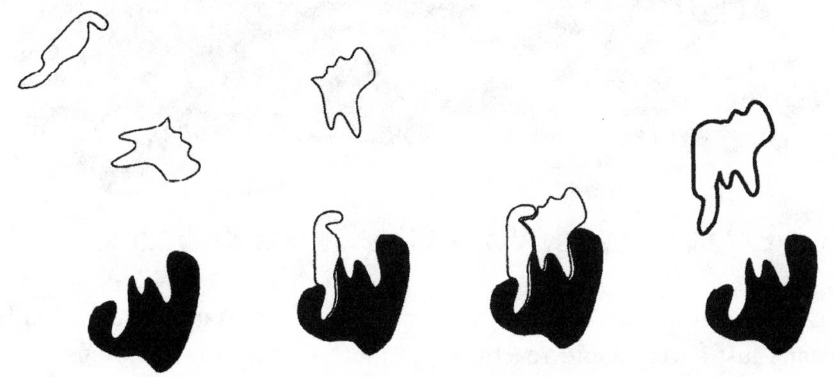

Figure 1.17.1. Some enzymes speed up reactions by holding the reactants together in the right orientation. The reaction is shown in four steps going from left to right. First the reactants (white) are floating freely in solution. Then they attach to the enzyme (black). Held together, they react. The finished product leaves the enzyme and the enzyme is free to catalyze another reaction.

For other catalysts, the mechanism is less well understood, and the basis for choosing a catalyst for a particular reaction is still something of an art. For instance, a copper wire will catalyze the decomposition of hydrogen peroxide much better than a similar strand of steel wool. This difference can be demonstrated satisfactorily by putting on safety glasses and then dropping a copper wire in a half cup (120 milliliters) of drugstore hydrogen peroxide and a few strands of steel wool in a similar hydrogen peroxide sample. Along the copper wire there will be bubbles as the hydrogen peroxide breaks down into oxygen gas and water.

Because of the many factors that influence reaction rate—concentration, temperature, catalyst—and the fact that the reactants come together randomly and not always in the right orientation or with the right energy, there is no straightforward recipe for predicting reaction rates. Chemists are making good progress, and maybe some day we will know enough to control reaction rates completely, but for right now, for most reactions, we are stuck with the speed limits that nature has posted.

## For Example: When Your Car Comes Knocking

The word *explosion* is often associated with devastation, but it really means just a very rapid reaction—so rapid that the surroundings don't have time to accommodate the change. In our modern world, explosions are as common as breathing. The internal combustion engine can generate some several thousand explosions per minute.

It seems odd that the internal combustion engine has only been around for a century or so while humankind has been making explosions for several thousand years. The explanation, of course, has to do with all the engineering that was necessary to harness the explosions, including how to make several explosions operate at once and have them all work together. In other words, it's all in the timing.

In an internal combustion engine, gasoline and air are mixed in a sealed cylinder that houses a movable piston. The piston moves up and compresses the gasoline-air mixture, which is then ignited with a spark. The resulting explosion pushes the piston down, and the downward motion moves a lever that turns the driveshaft. The explosions are timed so that when one piston is pushing down, part of its downward force is being used to move another piston up, seesaw fashion—or at least that's what happens when things are going right. One of the things that can go wrong is a condition called "engine knock," an imperfect explosion that can result in "pinging," or sudden loss of power. The cause of engine knock can be understood in terms of chemical reaction rates.

In a simple picture of an engine operating, it may appear that a piston moves up, compressing the gasoline-air mixture, until it reaches the top of its upward stroke, and then the spark plug fires, the gasoline explodes, and the piston moves down. But this simple picture neglects the fact that the burning of gasoline is a chemical reaction and like all other chemical reactions, including explosions, does not happen instantaneously. The explosion takes time to propagate though the gasoline-air mixture.

To account for the fact that the explosion takes time, the spark is usually set to occur right before the piston reaches the top of its upstroke. This way the piston will receive the shove from the explosion of gas just as it is ready to start moving down, and it will benefit from the full impact of the explosion. If the explosion were to start right at the top of the piston's stroke, then the piston would already be moving away as the full force of the exploding gases were moving toward it, and the impact of the force would be lessened. If the spark were to occur too far in advance of the piston, then the full explosion could meet the piston on its way up and work against the power stroke of the other pistons down the line.

The benefit of all this intricate timing can be destroyed, however, by unauthorized ignition. If something happens to cause the gasoline-air mixture to ignite before the spark plug has a chance to spark or causes the mixture to start burning at several places at once, the result will be shock waves. These uneven and unaccounted for pressures can seriously interfere with the functioning of the piston.

Temperature, as we've seen, is an important factor in reaction rates. Increasing the pressure, as you may recall from our discussion of the gas phase, can cause an increase in temperature. It is possible to heat the gasoline-air mixture to its ignition point by pressure alone, and, in fact, diesel engines are based on this concept. Diesel engines do not have a spark plug but rely on compression to ignite the fuel-air mixture. If this occurs in a nondiesel engine, however, it spells uncontrolled burning and trouble. The gasoline-air mixture has to be carefully regulated in a well-tuned engine to prevent knocking.

Another factor is fuel. If the gasoline is of a grade that is better able to withstand the temperatures and pressures within the cylinder without spontaneously igniting, then knock is less likely. A fuel with a high "octane" rating is formulated to resist burning until it is sparked.

The name *octane* denotes a compound with eight carbon nuclei in its structure. The *octane rating* of gasoline originally meant the ratio of a particular type of octane (there are several ways to arrange eight carbons in a chain) to a particular seven-carbon chain. Since its inception, however, the term has evolved to describe the burning characteristics of a fuel more than its exact composition. For instance, when it was found that certain lead compounds could improve the burning behavior of gas and thereby avoid knocking, lead additives were said to raise the "octane" level of the gasoline.

Fortunately, engine design is also a factor in preventing knock, so when it was found that lead in the fuel became lead in the environment, cars were redesigned to accommodate a lower octane-rated fuel. But shouldn't all cars routinely use a high-octane fuel anyway, just to be on the safe side? Not necessarily. Some cars run better with a lower octane fuel. The benefit from a high-octane fuel will depend on engine design, age, condition, location, tuning, and time of the year. Experience is probably your best guide. If your car is running well, don't knock it.

# Demonstration 18: Black-Light Chemistry, Paper-Clip Chemistry

*The persons who constitute the natural aristocracy are not found in the actual aristocracy, or, only on its edge: as the chemical energy of the spectrum is found to be greatest just outside of the spectrum.*

—Ralph Waldo Emerson, *Essays*, ca. 1840

In our catalog of the many factors that influence engine performance, one we did not list was intensity of sunlight because the reactions in engines take place where the sun doesn't shine. But sunlight does affect cars. The chemicals in car paint, tires, and plastic fixtures can be degraded by the sun. In fact, light can be a factor for many reactions. Light can also be a mysterious factor because many times it is not just light, but the type of light, that matters. To demonstrate the effects of light on a chemical reaction, revisit the demonstration of the last chapter in which we measured the rate of bleaching of an indicator solution.

Put on your safety glasses, then set up the cups with bleach as before, but this time expose one of the solutions to the black light suggested for purchase in the "Shopping List and Solutions." The black light should be moved to within an inch of the solution's surface for maximum effect. The light should shine directly on the surface, not through the glass or plastic of the container. The exposure should be for about fifteen minutes. Take care that the other solution is not exposed to the black light. Now add equal drops of indicator to the cups of bleach and time how long it takes for the color to fade. The reaction should run significantly faster for the solution that has been exposed to the black light because the bleach has been *photochemically degraded*; that is, the black light has caused the bleach to be chemically broken down into more reactive species.

Now, with safety glasses still in place, pour about a half inch (1 centimeter) of copper sulfate solution (made as directed in the "Shopping List and Solutions") into a disposable glass or plastic cup. Unbend a paper clip and allow one end to dangle in solution. Put one end of a galvanized nail into the solution, taking care not to touch the paper clip. Put one end of a length of aluminum wire into the solution, making certain it doesn't touch either of the other metals. Wait about an hour and then take the metals out and examine them. The paper clip should have a nice copper coating, the galvanized nail should be darker where it has touched the solution, and the aluminum wire should be unchanged. With the paper clip and the galvanized nail, electrons at the surface of the metal have been transferred to the copper ions in solution. Metallic copper formed when the ions accepted the electrons. Copper deposited on the paper clip and nail surface as iron in the paper clip and the nail ionized and dissolved. The electrons, however, have a harder time leaving aluminum. An exchange of electrons through solution or through metal is termed *electrochemistry*. We'll investigate more of these reactions—and their causes and consequences—next.

# CHAPTER 18

## Electrons and Photons: Turning on the Light

*The lightning that preceded it*
*Struck no one but myself,*
*But I would not exchange the bolt*
*For all the rest of life.*
*Indebtedness to oxygen*
*The chemist may repay,*
*But not the obligation*
*To electricity.*

—Emily Dickinson, *Poems,* third series, ca. 1850

What do fireworks and fall foliage have in common? Photochemistry: chemistry in which light is a product or reactant. In fireworks light is a product and evident to the eye. In producing fall foliage, light is a reactant: placing a paper bag over a clump of deciduous tree leaves will prevent them from turning colors. Light is a reactant in fall's most famous reaction.

What do batteries and bumpers have in common? Electrochemistry: chemistry in which electricity is a product or reactant. Batteries produce electricity via chemical reactions and create enough current to light up a lightbulb. Electroplating employs a stream of electrons to reduce a metal, such as chromium, onto a surface, such as the bumper of a 1956 Crown Victoria. What do photochemistry and electrochemistry have in common? Both are types of reactions that involve small particles that are not atoms but that do chemistry. In electrochemistry the particles are electrons, with which we are now familiar. In photochemistry the small particles are photons, which beg a bit of explanation.

We begin with the nature of light. Light can be characterized as an *oscillating electromagnetic field*, a term that wants dissecting on its own. In the physical sciences, a *field* is an effect that extends out over a region of space. For instance, if there is a bonfire in the middle of a back yard it will create a temperature field: the temperature will be highest where the fire is centered, then extend out and gradually diminish with distance from the fire. An *oscillating field* is a field that varies regularly with time. An oscillating fan blowing over a block of ice could serve as an example of an oscillating temperature field. An *oscillating electromagnetic field* is an oscillating electric field accompanied by an oscillating magnetic field.

You can demonstrate the relationship between electricity and magnetism by using a nail, a length of wire, a D-cell 1.5-volt battery, and a battery holder (the last item being optional). The wire is wrapped around the nail in a series of tight coils and then the ends are attached to (or held to) the ends of the battery. As a result, the nail becomes an electromagnet that will pick up paper clips. The magnet can be switched on and off by connecting and disconnecting the battery. The wire may become hot, so don't keep it attached longer than necessary. Never store a battery with wires attached, as they may come in contact and become hot enough to ignite flammable material.

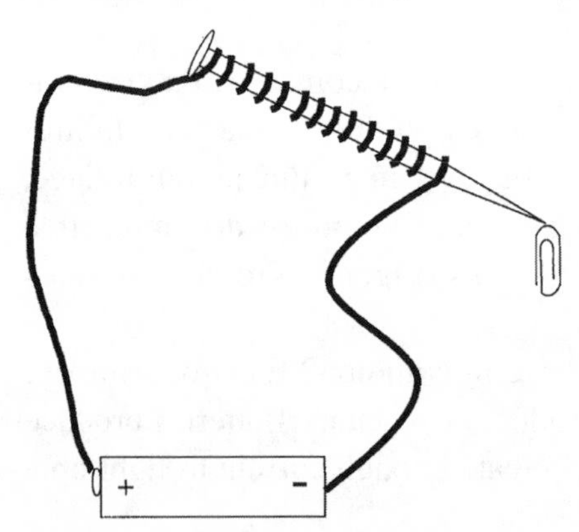

Figure 1.18.1. A homemade electromagnet can pick up paper clips.

Why does a moving current create a magnetic field? To be perfectly honest, no one has completely explained it yet, so we won't try here. Suffice it to say that the effect is well established and makes for some interesting chemistry, too.

Why would an oscillating electromagnetic field have anything to do with chemistry? As may be recalled from our discussion of atomic structure, the atom itself has an electric field, as we illustrated with our "Water Witch" demonstration. The electron has a negative charge and the proton has a positive charge, and the two are separated, so there is a field—an attraction—between the electron and the proton. The molecule may absorb energy from the oscillating electromagnetic field of light as the electrons are pushed and pulled from their normal positions, as a child in a swing will absorb energy from the parent as the child is pushed from the equilibrium position in the swing. All of chemistry boils down to a rearrangement of electrons, so electrons pushed around in this manner may find themselves in new situations, which may result in new materials.

But why did we say that light *may* interact with atoms? If light is an oscillating electromagnetic field and the atom contains an electric field, shouldn't light *always* interact with atoms? The answer may be found in another question you may have asked yourself during the demonstration for this chapter: Do I have to buy a black-light lamp? Those things are expensive! Can't I just use a regular lightbulb or just set the solution out in the sun?

The answer is that the black light provides extra ultraviolet light (UV light), the type of light needed for the reaction. The action of UV light also explains why chlorine levels in swimming pools have to be monitored. They cannot be counted on to change by the same amount every day because there is not the same amount of sunlight every day. But if the sun gives off ultraviolet radiation, why don't we just set the reaction outside and save the expense of a black-light lamp? The answer is that ultraviolet light isn't the only light emitted by the sun, and the presence of other types of light would confuse our experimental results. The sun also gives off visible light and infrared light, and the latter is capable of heating the solution. As we mentioned in our discussion of chemical kinetics, heat can speed up a reaction. So if we set our solution in the sun, we would not be certain if it were the UV light or the heat causing the breakdown of the bleach. The type of light needed for a reaction is very specific because the type of electric field surrounding a molecule is very specific. The field depends on the nature of the elements that make up the molecule and their positions in the molecule.

The energy carried by light depends on the type of light. The type of light that would heat our solution if we put it in the sun (infrared) is relatively low-energy light; the visible light that our eyes respond to is a moderate-energy light; and the UV radiation that caused the chemical reaction is relatively high-energy light. UV is the type of light that can cause skin damage if we stay out in the sun too long. The light required for a chemical reaction can be so specific that the light can be thought of as a stream of energy packets, each particle causing one chemical change. These energy packets are called photons, and they can be counted, just as we count electrons, molecules, and atoms.

How can light be a particle *and* a wave? This question has confounded generations of chemistry and physics students, especially because the answer is that it is neither—and it is both. Light can be modeled as a particle when the particle model is convenient, and it can be modeled as a wave when the wave model is convenient. Neither model captures the true essence of light, just one aspect.

An analogy might be found in the common house cat. Is a house cat a domestic animal or a wild animal? A cat behaves as a domestic animal when it comes when called for food or for affection, but it behaves as a wild animal as soon as a spider has the bad fortune to scuttle across its path. So is the cat domestic or wild? Sometimes one description works, sometimes the other. The cat is neither and it is both. Our light is sometimes described well as a particle and sometimes as a wave.

We've already seen that light is able to induce chemical reactions, but how can we show that chemical reactions are able to produce light? Probably the easiest way is to go to a county fair or a circus or a novelty store and buy a glow stick. These plastic-encased elongated tubes are activated by bending and/or twisting until a membrane separating two solutions is broken and the solutions mix and react. These solutions contain materials that phosphoresce when they react, and the observed effect can be quite pleasing. Another example can be found in fireworks. The metal salts in the fireworks are heated by a gunpowder explosion to the point where the electrons are disturbed in their orbitals. As the atoms relax back to their more comfortable, stable state, they emit excess energy in the form of light.

Photochemistry, chemistry caused by light, is responsible for many of the atmospheric reactions that determine the quality of our air. Photochemical smog is the name given to the red-brown haze that can form over cities under certain weather conditions. Nitrogen (as $N_2$) and oxygen (as $O_2$) are natural components of the atmosphere and can form NO in lightning strikes. NO reacts with $O_2$ to form $NO_2$, which dissolves in rainwater and falls to the ground. Here the nitrogen can be latched onto by bacteria and, through a series of reactions, make proteins for plant and animal life forms. Unfortunately, there is another not-so-natural source of atmospheric $NO_2$, and that is from the cylinders of automobiles. The resulting $NO_2$ would not be a problem if it were present in small amounts, but in large quantities it can cause damage. Released into the atmosphere, $NO_2$ can absorb UV light from the sun and break down into NO and O again. The difficulty is that O can then react with $O_2$ to form $O_3$, commonly called *ozone.* While ozone in the upper atmosphere is a good thing—it shields Earth's surface from the full force of the sun—down at street level, it isn't so good. Ozone damages rubber, plastic, plants, and animals and reacts with car exhaust to form irritating pollutants.

So energy from light causes some chemical reactions. But why do some reactions give off light? Because a chemical reaction consists of a rearrangement of electrons, and depending on the reagents and products, these electrons may find themselves in product molecular orbitals that are not at the lowest energy. Just like the electrons in our fireworks, as they work their way back to a stable ground state, they may emit light.

There is another way in which electrons can be rearranged in a chemical reaction, and that is through a wire. Electrochemistry is redox chemistry wherein the site for oxidation is separated from the site for reduction. Electrochemical setups basically come in two flavors: electrolytic and voltaic (also known as galvanic) cells. Voltaic cells are cells that produce electricity, so a battery would be classed as a voltaic cell. The principles that drive voltaic cells are the same that drive all other chemical reactions, except the electrons are exchanged though a wire rather than direct contact. The reactions are redox reactions (which is why they produce an electron current); the reactions obey the laws of thermodynamics and move toward equilibrium (which is why batteries run down); and the reactions have defined rates (which is why some batteries have to be warmed to room temperature before they produce optimum output).

But electrochemical reactions have to obey the laws of electronics as well as the laws of chemistry, so even if the solutions are connected by a wire, the circuit has to be completed or the electrons will travel only so far and no farther. Contact is normally achieved by an ion-containing solution called an *electrolyte*. Because ions are charged particles that move in the electrolyte, they are able to conduct charge from the oxidation site to the reduction site as the electrons move through the wire. In batteries that you buy from a store, the electrolyte is usually in the form of a moist paste rather than a liquid. But in high current-producing car batteries, the electrolyte is usually sulfuric acid.

A lemon is acidic, so lemon juice actually makes a fair electrolyte, too. While wearing safety glasses, stick a zinc-coated nail (a galvanized nail) in one end of a lemon and a penny into a narrow slit on the other end. Now measure the voltage between the nail and the penny with a hardware store voltmeter, attaching the negative lead to the galvanized nail and the positive lead to the copper. The actual voltage measured will depend on the lemon and other factors.

The reaction in the lemon battery is the reduction of hydrogen ions to hydrogen gas (lemon juice is acidic) and the oxidation of zinc into zinc ions—the thermodynamically more stable state. Because the reaction is moving toward a more stable state, it can produce electricity as a voltaic cell. An electrolytic cell is the antithesis of the voltaic cell. In the voltaic cell, a chemical reaction is used to produce electricity. In an electrolytic cell, electricity is used to produce chemistry. A demonstration electrolytic cell can be set up as follows.

Make certain you are still wearing safety glasses and then take your copper-coated iron nail from the first demonstration and dangle it in a glass with about an inch of the freshwater aquarium pH-lowering solution suggested for purchase in the "Shopping List and Solutions." Hook it via a wire to the positive end of a D-cell battery in a battery holder. Add a companion iron nail to the solution and hook it to the negative end of the same battery. The copper will gradually leave the first iron nail and plate out on the second. This process, called electroplating, has been used to add metal coating to everything from furniture to jewelry. The purpose of electroplating can be decorative, but is usually more practical. Electroplating can successfully seal a reactive metal, such as iron or iron alloy, from the corrosive affects of moisture. Why is moisture damaging to iron? One would think that a steel bridge that can withstand the weight of tons of traffic every day would be able to withstand a few drops of water. But water drops on iron form an electrochemical cell just as an iron wire connecting two solutions. In the rust-forming reaction, iron is oxidized to iron oxide, which is rust, and reduced oxygen from the air provides the oxide. To demonstrate that this reaction happens at two different sites—one for oxidation and one for reduction—unwind a paper clip and set it half in and half out of a glass of vinegar for a couple of days. Then take it out and look at it under a magnifying glass. You should see the rust piled up at one specific location and below this spot a small clean-looking area. Rust forms above the solution because the reaction requires oxygen.

So now we have demonstrated that fireworks and fall foliage have something in common, and batteries and bumpers have something in common. But in this chapter, we have grouped all four together, so what do fireworks, fall foliage, batteries, and bumpers *all* have in common? The answer? Solar energy.

## For Example: Photon Meets Electron—Solar Energy

Basking lizards and sunflowers were taking advantage of solar energy long before intelligent life came along. Broadly speaking, we could say that all life is based on solar energy. But we have added a little technology over the years. To begin with, we have borrowed from the lizard idea and developed solar collectors that absorb energy from the sun and store it. We have also developed photovoltaic cells that turn light from the sun into an electric current. Photovoltaic cells are being used successfully to power small devices such as calculators and even for large-scale energy production in areas where the sunshine is abundant. Scientists are also working on other creative technologies that will tap into the abundant energy from the sun. One of these is use of solar energy for the electrolytic degradation of pollutants.

Some of the more difficult pollutants to deal with are the organic pollutants that derive from fertilizers and insecticides. Organic compounds are the types of materials that make up oils, waxes, and plastics and can be very resistant to attack by water, simple salts, or even acids and bases. However, organic materials can be broken up by putting them in an electrolytic cell and driving electrons through them or removing electrons from them. But the electrons have to come from somewhere. If the current is produced in an oil-burning plant, then the process of breaking down pollutants creates pollutants, and we're back to square one. Fortunately, there are materials such as titanium oxide particles.

Titanium dioxide is used in paints because of its pure white color. When the sun is out, the titanium oxide particles can absorb the solar radiation and use this energy to promote its electrons to an excited, more reactive state. These electrons then act like tiny electron sources on diminutive batteries to break down organic pollutants, molecule by molecule. So as in the Aesop tale—the quarrel between the sun and the wind as to which was the stronger—that which resists force can often be conquered by gentle persuasion. The warm, welcome light of the sun.

# PART 2

# INTRODUCTION

## Playing the Tunes

*"I am happy," said M. Waldman, "to have gained a disciple; and if your application equals your ability, I have no doubt of your success. Chemistry is that branch of natural philosophy in which the greatest improvements have been and may be made: it is on that account that I have made it my peculiar study."*

—Mary Shelley, *Frankenstein*, ca. 1820

We have covered a great deal of territory. We discussed the nature of chemistry, the structure of atoms and chemical compounds, and the properties of elements as reflected in the periodic table. We explored fundamental classes of chemical reactions such as redox reactions, acid-base reactions, and displacement reactions. We laid out the theory that describes the nature of the chemical bond and the principles of chemical reactions. We discussed the practical considerations of intermolecular forces and concentration and considered the rarefied properties and reactions of gases. We contrasted the slippery properties of solutions with the concrete properties of solids. We delved into thermodynamics and explored phase equilibrium. We examined the important concept of chemical equilibrium, and we saw how the principles of thermodynamics predicted the observed colligative properties of solutions. We looked at the precepts of chemical kinetics, and we wrapped up our survey of the principles of chemistry with a look at chemistry induced by electricity and light.

Whew!

A lot of head scratching and page turning, but now we are ready for a payoff: seeing how these basic principles come together in some of the remarkable specialized fields of chemistry—organic chemistry, inorganic chemistry, biochemistry, and analytical chemistry—and a look ahead to an intriguing future.

Before we begin, let us indulge in a bit of an overview.

Organic chemistry and inorganic chemistry are not *applications* of chemistry so much as they are *explorations* of chemistry, but explorations with specific goals in mind. The organic chemist is interested in materials with a backbone of covalently bonded carbon and hydrogen: hydrocarbons and their derivatives. The objective of organic chemistry is to synthesize and understand the properties and reactions of these interconnected, looping chains of carbon in all their infinite variety. These materials are found throughout our oily, waxy world and in the materials from which our bodies are made.

The inorganic chemist investigates the materials made from the other one hundred and some elements on the periodic table, an enormous undertaking. But while these compounds are rich and intricate in their behaviors, they do not form long, interconnected chains of the same element like carbon does. Their diversity is in their elemental composition. Inorganic materials form the rock and salty solutions of Earth, the planets, and all the materials of the stars. Inorganic chemists devise syntheses for semiconductors, superconductors, alloys, and many other new materials.

The biochemist studies the identity, properties, and interactions of the materials that are the foundation of life. This important, exceedingly complicated endeavor draws from all disciplines of chemistry and employs all the principles of chemistry to understand the chemical reactions that drive the rhythms of life.

The analytical chemist draws from all of the above. The objective of the analytical chemist is to identify and quantify materials—be they organic, inorganic, or biochemical—sometimes by separating them from the soup and sometimes as they sit in their murky matrix. Each chemist must have some command of the art of analyzing chemicals for their composition and concentration. The analytical chemist finds employment as an integral part of academic, industrial, and governmental laboratories. We will gain some appreciation for the virtuosity of the analytical chemist by following some of the adventures of a special breed of analytical chemist—the forensic chemist—during the course of an interesting day.

And so, to the lab. . . .

# Organic Chemistry Demonstration: Aspirin to Acid

*It was in their undergraduate days, however, in the midst of their profoundest plunges into the mysteries of organic chemistry, that Doris Van Benschoten entered into their lives.*

—Jack London, *Moon-Face and Other Stories*, ca. 1905

*When I had succeeded in dissolving the hydrocarbon which I was at work at, I came back to our problem of the Sholtos, and thought the whole matter out again.*

—Arthur Conan Doyle, *Sign of the Four*, ca. 1890

Owing to several unique properties of carbon, such as its ability to form long chains, organic compounds—carbon compounds—come in virtually infinite variety. Organic chemists spend their time studying the ways that carbon reacts with itself and other compounds and use this information to design *syntheses*, stepwise procedures by which a specific compound can be produced from a given starting material. Using their knowledge of reactivity of carbon, organic chemists have designed syntheses for nearly all the marvelous organic materials that we enjoy, from plastics to medicines. However, this pliability on the part of organic molecules also represents the challenge of organic chemistry. Synthetic schemes have to be carefully considered and may involve many steps, each of which may necessitate careful controls. The following demonstration provides a small taste of the steps that may be involved in such a synthesis.

In this demonstration, we first isolate the active ingredient of the modern form of aspirin, acetylsalicylic acid, using a time-honored organic chemistry procedure called *extraction*. The preparation of purple-cabbage indicator is a type of organic extraction: the organic purple dye is extracted from the cabbage with water. Here we use alcohol to extract acetylsalicylic acid from aspirin.

We will then convert our acetylsalicylic acid to acetic acid with a step-by-step organic synthesis. The equation for the reaction is given below, and figure 2.1.1 shows the structure of the molecules in the reaction. The lines between the symbols for the elements represent the bonds. Note that some of these are double bonds. Carbon is a very versatile element.

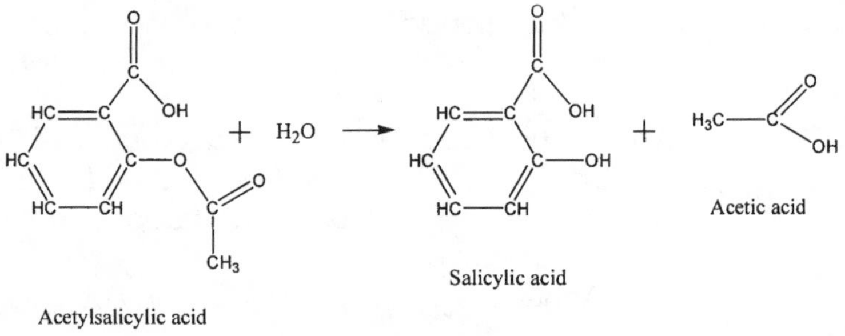

Acetylsalicylic acid + $H_2O$ → Salicylic acid + Acetic acid

$$CH_3CO_2C_6H_4CO_2H + H_2O \rightarrow C_6H_4CO_2HOH + CH_3CO_2H$$

Figure 2.1.1. Acetylsalicylic acid, the active ingredient in aspirin, can be converted to salicylic acid and acetic acid, which is a principle component of vinegar.

Organic chemistry is a rigorous, demanding discipline, but with a little aspirin it can be less of a pain.

## Extraction and Isolation of Acetylsalicylic Acid

Place ten to fifteen aspirin, plain or buffered, into a glass container. With your safety glasses on, pour in a half cup (120 milliliters) of rubbing alcohol; 70 percent isopropyl alcohol works just fine. This amount should be enough to cover the tablets with some to spare. It doesn't take much solvent to extract the acetylsalicylic acid from the tablets. Gently warm the container in a microwave on half (50%) power for thirty seconds, then remove it from the microwave oven. The solvent should be warm but not boiling. The warm solvent will extract the aspirin compound from the tablet and leave the starch binder and other components undissolved. Starch is a carbon-containing polymer. The long chains of the polymer will not dissolve in the alcohol-water solvent. If you have an old spoon or fork, you may want to gently crush the residual material to maximize the amount of acetylsalicylic acid that is extracted. This process takes about fifteen to twenty minutes, but your patience should be rewarded.

Take a paper towel and spread it over the top of a large glass. Push it down slightly so that it resembles a funnel. Carefully pour the aspirin and alcohol solution through the paper towel filter. This will collect the tablet residue in the towel while allowing the clear liquid, called the *mother liquor*, to trickle to the bottom of the glass. Once all of the solution is filtered, put on your rubber gloves and dispose of the paper towel and tablet residue. Try not to touch the moist portion of the paper towel, as the solvent is laden with acetylsalicylic acid, which could remain on your skin after the solvent evaporates. You may want to wash your hands at this point, although aspirin is not a dangerous compound in the miniscule amounts that may be on your hands. The acetylsalicylic acid is now dissolved in the alcohol in the glass.

Run a small stream of cold tap water and fill the glass containing the mother liquor with the cold tap water until it is about three-quarters full. If your alcohol extraction was successful, there should be many very small white flat crystals of acetylsalicylic acid filling the mixture. Acetylsalicylic acid is somewhat nonpolar and dissolves fairly easily in a somewhat nonpolar solvent like isopropyl alcohol. But as soon as the very polar water is added, the acetylsalicylic acid is forced out of solution by its decreased solubility and it forms crystals. These are not large single crystals but tend to look like white soap flakes and may appear gooey until they are dried.

It is always impressive to observe the volume of crystals formed, especially when compared to the original tablet volume. Set up another paper towel filter in a large glass and collect the acetylsalicylic acid crystals in the towel by pouring the alcohol-water-crystal mixture through the towel. Rinse the crystals out of the first glass using small amounts of cold tap water, and pour those small portions through the filter. This will also help rinse off the crystals in the paper towel. Keep the crystals in the paper towel and dispose of the used solvent that dripped through.

It is possible at this point to go on to the next procedure, but if you have the time, spread the paper towel with the crystals out on a countertop for air drying, which may take overnight. Make certain the crystals are protected from children and pets, and vice versa. Strong breezes would also be bad because the dry crystals are light and fluffy. Observe that there is no aroma associated with these crystals once they are dried, not even that of isopropyl alcohol. Once the crystals are dry, you can observe how light they are by picking up the paper towel. Most of the original weight of the wet crystals was solvent.

## Acid-Catalyzed Hydrolysis to Salicylic Acid and Acetic Acid

If you let your crystals dry overnight, then find your safety glasses again and put them on. Take about a quarter of either the wet or dry acetylsalicylic acid crystals and place them into a glass container that can be used for warming. The dried crystals are very fluffy and should be handled carefully so they do not scatter. Take the bottle of aquarium pH-lowering solution recommended for purchase in the "Shopping List and Solutions," and add it dropwise until the entire sample of acetylsalicylic acid is completely covered. The pH-lowering solution is sulfuric acid that is not very concentrated and perfect for this reaction. It behaves as a catalyst in that it speeds up the reaction and is not consumed during the reaction. Be extra careful during this phase of the experiment: dilute sulfuric acid will remain at the end of the experiment, and even dilute acid should always be treated with cautious respect.

Gently warm this mixture for about fifteen seconds in a microwave oven on low power (50% power). Remove the container from the microwave and use your hands to waft any aroma from the top of the container toward your nose. At the end of the reaction, part of the acetylsalicylic acid will separate from that molecule because of the acidic environment. The part that leaves becomes acetic acid, the active principle of vinegar, and the aroma will eventually be detected by your nose. Once the acetic acid has evaporated, the crystal residue will be the simpler analgesic, salicylic acid. All solids can be disposed of in the trash, and all liquids can be rinsed carefully down the toilet.

The reaction you have just executed is an organic reaction known as an *acid-catalyzed hydrolysis.* The suffix *-lysis* means "to cut," and *hydrolysis* means "to cut with water." In this reaction, acetylsalicylic acid is "cut" with water to form salicylic acid and acetic acid, as shown in figure 2.1.1.

# CHAPTER 1

## Simply Organic

*On the other hand, I used to find Paul Tichlorne plunged as deeply into the study of light polarization, diffraction, and interference, single and double refraction, and all manner of strange organic compounds.*

—Jack London, *Moon-Face and Other Stories*, ca. 1905

*Even while we talk some chemist at Columbia*
*Is stealthily contriving wool from jute.*

—Robert Frost, *Build Soil*, ca. 1950

It is instructive to look around at the number of things composed primarily of organic materials—that is, derived from the covalently bonded carbon-hydrogen compounds called hydrocarbons. From asphalt to aspirin, our world is awash in hydrocarbon derivatives.

These materials include plastics, which nowadays make up part of everything from automobiles to office supplies, to farm implements, to clothes. Any food, from steak to French fries and sugar to sardines, is organic, which brings us to the biggest category: all living things. Plants, animals, fungi, and bacteria. Our fuels—gasoline, natural gas, propane, and butane—are organic because they derive from once-living things, such as trees, dinosaurs, and fish. Petroleum products range from asphalt to acetone. Coal is a lump of organic material. Our soaps are organic, as are our shampoos, whether they are specially touted as such or not. Makeup? Organic. Toothpaste? Deodorant? Mostly an organic base with a few active ingredients thrown in. Our medicines? Organic, or if the active ingredient isn't organic, the starch carrier or gelatin capsule is. Any flavored drink may be mostly water, but the colorings and flavorings, either natural or from coal tar, are organic. Clothing, natural or synthetic, is made up of organic fibers. Many times the word *organic* is taken to mean "produced naturally rather than manufactured," but in truth, all hydrocarbon derivatives are organic, regardless of how they came into being. But there can be no doubt that the first organic compounds were natural. The body wearing the clothes is organic, too.

Why are so many materials derived from hydrocarbons? Actually, the mystery does not go that deep. Deep into Earth, that is. When we compare the mass of Earth, including its inorganic core, with the total mass of organic compounds, organics are not that common after all. Carbon does not rank in the top eight elements that compose our planet (which are, in order of abundance, iron, oxygen, silicon, magnesium, nickel, sulfur, calcium, and aluminum) or even among the top eight elements making up the Earth's crust (which are, in order of abundance, oxygen, silicon, aluminum, iron, magnesium, calcium, potassium, and sodium). But the carbon we have is concentrated at the very upper edge of Earth's crust because of carbon's small atomic mass. During the upheavals over the eons of Earth's existence, the lighter elements have shifted to the top, like peppercorns in a sea of peas. Once in the crust, carbon was able to form the largest variety of compounds primarily because of its unique ability to form long, stable chains of bonds, a talent called *concatenation.* The strength of the stable bonds of organic compounds is evident when those bonds are broken and reformed: organics make excellent fuels.

Carbon compounds can grow into beautifully complex and intricate chains because carbon can form four separate bonds with other elements, and it can form stable bonds with itself. There are, of course, other elements that can self-bond, such as $N_2$ and $O_2$, and there are still other elements that can form chains, such as a cyclic form of sulfur, $S_8$. But this ability to form so many varieties and lengths of chains, and remain pliable or fluid while doing so, is a gift that is uniquely carbon's. Carbon compounds can also have a three-dimensional character that is necessary to build our three-dimensional world. Methane, the most basic organic molecule, is formed from one carbon and four hydrogen atoms. Carbon sits at the center of this molecule, and the four hydrogen nuclei extend outwards to the four corners of a regular tetrahedron, as shown in figure 2.1.2. This type of drawing is called a *projection* drawing because the solid and dashed wedges indicate which bonds would project into and out of the plane of the page in the true three-dimensional structure.

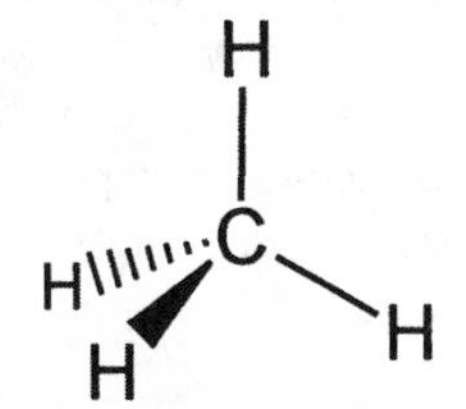


Figure 2.1.2. The structure of methane, $CH_4$. The straight-line bonds indicate hydrogens in the plane of the paper. The solid-wedge bond indicates a hydrogen coming out of the plane of the paper and the dashed-wedge bond indicates a hydrogen on the back side of the plane of the paper.

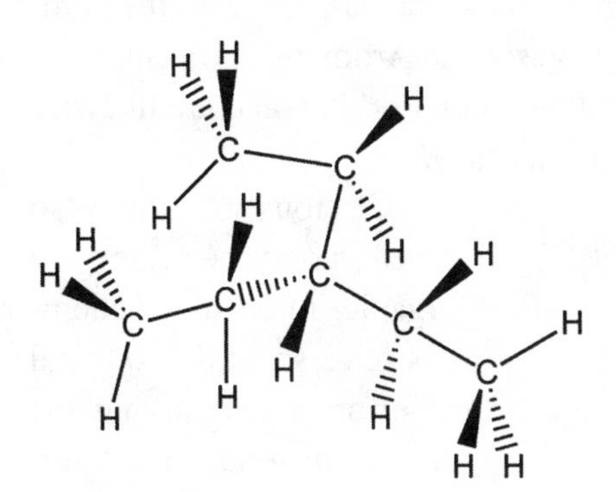

Figure 2.1.3. A seven-carbon chain twisting about its bonds.

From the angles of this one molecule, it may be imagined how intricate scaffolding may be built, twisting and turning into varied filigreed shapes. A relatively minor example, one involving only seven carbons, heptane, is shown in figure 2.1.3.

And carbon is even cleverer than this. Carbon can form double and triple bonds with itself, as shown in figure 2.1.4. Multiple bonds result in triangular and linear geometries, which extend the architectural choices. When a compound has all single bonds, it is said to be saturated. In other words, methane shown in figure 2.1.2, and heptane in figure 2.1.3, would be said to be saturated, while the structures in figure 2.1.4 would be said to be unsaturated. To gain an appreciation of the dif-

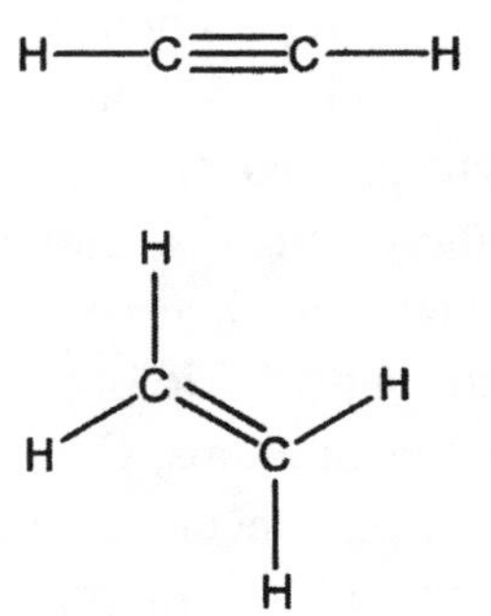

Figure 2.1.4. Double and triple bonds in carbon molecules. The double-bonded compound is commonly called ethylene, and the triple-bonded compound is commonly called acetylene.

ference that this simple change makes, think about all the fuss made over saturated fats versus unsaturated fats. What a difference a double bond makes! We'll revisit saturated fats when we discuss the role of fats in biochemistry.

As a result of this bonding virtuosity, carbon can also form circular, daisy chain–like *cyclic* compounds and even join several cycles together, as shown in figure 2.1.5. Beefed up by honeycombed rings, these types of compounds are robust enough to withstand the pressures of life and are found in many biologically essential materials. Once formed into a circle, carbon nuclei sometimes share electrons in a large molecular orbital that extends over the entire ring. The bond that results is given the descriptor *aromatic* because it is found in many odoriferous organic molecules, such as benzene, the compound that lends a pungent aroma to gasoline. Aromatic molecules are especially stable, owing to their shared molecular orbital, and are likewise found in many biologically important materials.

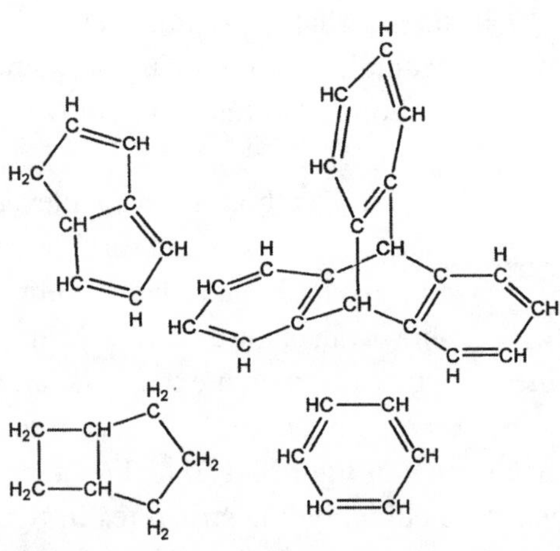

Figure 2.1.5. Some hydrocarbon structures that involve rings.

Obviously, with so much variety possible with organic molecules, chemists have spent a good deal of effort in classifying the various materials into families of compounds. The method that has evolved is to group compounds into families defined by the presence of certain identifying structures called *functional groups.* As examples, we will look at four families of organics that contain oxygen in their functional group: alcohols, ethers, organic acids, and esters.

For alcohols, the functional group, or defining structure, is OH. Alcohols all incorporate an OH group, but the nature of the rest of the molecule determines the unique properties of that alcohol. Ethanol is the alcohol present in alcoholic beverages, isopropyl alcohol is the alcohol of rubbing alcohol, and methanol is sometimes used for fuel—and they are not interchangeable. While it may be acceptable for adults to imbibe small amounts of ethanol on special occasions, it is never advisable to drink isopropyl alcohol or methanol, as these alcohols can cause bodily damage or death. Methanol, which can contaminate bathtub gin, is notorious for causing blindness or brain damage. Figure 2.1.6 shows these three structures. Notice how a small structural change can have a big impact on chemical behavior.

The defining arrangement for the *ether* functional group is oxygen between two carbons. (See figure 2.1.7.) One particular ether, diethyl ether, is known for its soporific effects on the mammalian brain, an effect that was widely appreciated by surgical patients when it was perfected as an anesthetic. Organic acids, shown in figure 2.1.7, also contain oxygen in their defining functional group, which is $CO_2H$. Acetic acid, which we have used extensively in our demonstrations in the form of vinegar, is an organic acid and has the formula $CH_3CO_2H$.

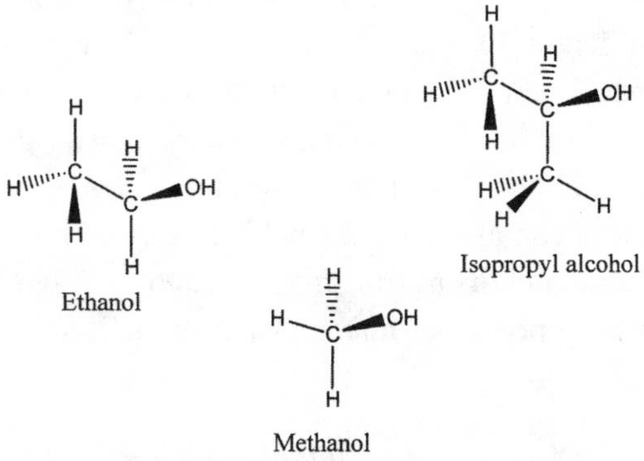

Figure 2.1.6. Three different alcohols with distinctly different properties.

Examples of the third type of functional group shown in figure 2.1.7, esters, are derived from organic acids. In an ester, the hydrogen in $CO_2H$ is replaced with an organic group. Some esters are noted for their pleasant fruity aromas and have been used to simulate the odor, and hence the flavor, of bananas, pears, and pineapple. Modern aspirin, acetylsalicylic acid, is an ester. In figure 2.1.7, we show an example of each: an ether, an acid, and an ester.

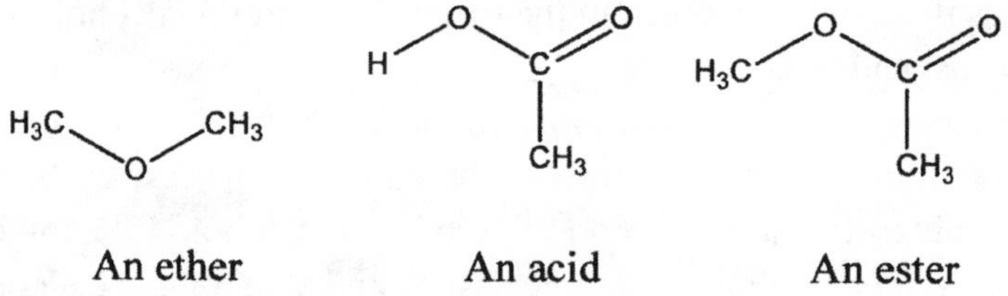

$H_3C$ O $CH_3$ | H O C O $CH_3$ | $H_3C$ O C O $CH_3$

**An ether** **An acid** **An ester**

Figure 2.1.7. An ether, an acid, and an ester.

Carbon compounds may also incorporate atoms of other elements, called *heteroatoms*, such as nitrogen, sulfur, chlorine, bromine, or others. But whatever the identity of the bonded species, carbons will always sport four bonds, be they single, double, or triple bonds. When a carbon has four single bonds, an interesting property can be created: handedness.

Handedness in a molecule is similar to handedness in a human. Humans are configured with right hands and left hands, and these hands are distinctly different. One cannot be mistaken for the other. They match point by point when they are pressed palm to palm or held up to a mirror, but when you try to stack one hand on the top of the other—the palm of one hand on the back of the other—you find your thumbs going in two different directions. Molecules can also have a handedness. A carbon with four different attachments can match its mirror image face to face, but lain on top of its mirror image, it may no longer match. An example is shown in figure 2.1.8.

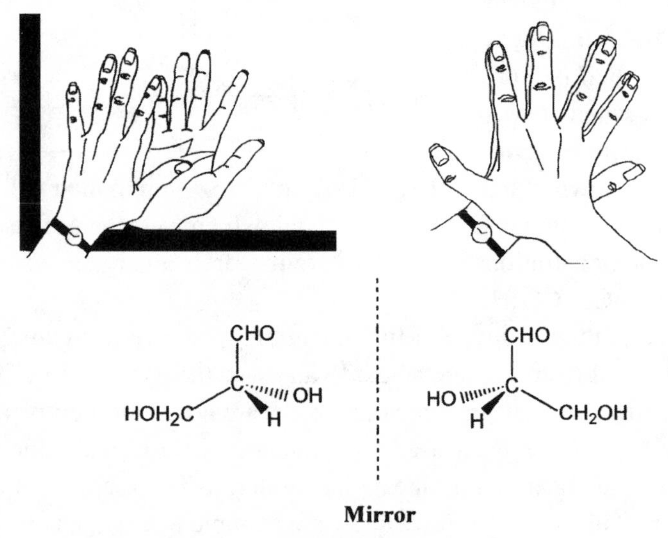


Figure 2.1.8. Chiral hands, chiral molecules.

As may be recalled from our discussion of the law of definite proportion, *isomers* are molecules that have been built from the same number and type of atoms but arranged in a different order. We cited, as examples, fulminic acid, cyanic acid, and isocyanic acid: HCNO, HOCN, and HNCO, respectively. We saw that this simple rearrangement of elements made the first explosive, the second a poison, and the third a pacific participant in several, more constructive, organic syntheses. Isomers that differ only by being mirror images of each other are termed *chiral isomers* (pronounced *kiral*, with a hard *c* sound, the way *chemist* is pronounced *kemist*).

Chiral twins have different properties from one another. For example, one form of carvone, a chiral organic molecule, smells of spearmint, while its mirror image smells of caraway. One form of limonene, a chiral organic molecule, smells of lemon, while its mirror image smells of oranges. To understand why this subtle difference in shape should make such a large difference in properties, try shaking hands with yourself or putting your right shoe on your left foot. It just doesn't work. The shoe doesn't fit. In many biological systems, if the molecule doesn't fit, it doesn't work.

These mirror-image molecules can also be distinguished from each other by the way they interact with polarized light. Polarized light is light of a limited range of orientations. In the sunglasses business, light of all orientations is called "glare." To experience the difference between glare and polarized light, flip on and off a pair of polarized sunglasses while looking at sunshine reflected off a shiny surface. Polarized sunglasses only allow through light of a limited range of orientations, or polarization, and thereby cut down on the glare. When light encounters a chiral molecule, the light may experience a twist from the electric field of the unsymmetrical molecule, as a car will turn if the road banks one way or the other. If there is a large number of such molecules interacting with the light, the effect can be substantial enough to be measured.

Using *l* to stand for *levorotatory* and *d* to stand for *dextrorotatory* (which is the chiral molecule equivalent of left-handed or right-handed), chiral molecules can be sorted into *l* and *d* categories depending on how they interact with polarized light. Many bodily molecules are chiral, including the sugars that are building blocks in DNA and the amino acids that make up proteins. As it turns out, nature often displays a preference for a particular chiral form of amino acids (see page 305). The mystery writer Dorothy Sayers used this idea in a 1930 story, *The Documents in the Case*, in which the victim is first assumed to have died of mushroom poisoning.[1] Closer examination of the toxins, however, reveal they are of the wrong chiral form and therefore of synthetic, not natural, origin.

The origin of this preference in nature is not known but probably has the same explanation as the British preference for driving on the left-hand side of the road: it's just the way things started out and it has always stayed the same. But not only can different mirror-image isomers cause different sensations of smell and taste, exposure to the wrong isomer of a biological compound can be dangerous. A notorious case of trouble caused by exposure to the wrong isomer is the case of Thalidomide, a sedative drug that was manufactured as a mixture of its two mirror-image forms. Unfortunately, the form that did not act as a sedative instead acted as a *teratogen*, a compound that causes birth defects. Because of such differences in activity, and also to lower the necessary dosage, it is desirable to produce medicines in a single isomeric form when possible. A good deal of research effort is currently being expended toward this goal.

Because the mirror-image isomers of chiral compounds are essentially separate materials, the number of chemically different organic compounds becomes all the more impressive. The agency that has assumed the responsibility for tracking such things in the United States, the American Chemical Society, now numbers the known organic chemicals at about twenty million, and more are discovered nearly every day. Obviously, finding names for all these progeny is a monumental task. Fortunately, the task, undertaken by the International Union of Pure and Applied Chemistry, has been made systematic.

Organic compounds are named for their characteristic features, just as a skinny redhead might be nicknamed Slim or Red. The alcohols described previously are named alcohols because they have the characteristic OH group. The other part of the name—methyl, ethyl, isopropyl—refers to the type of chain to which the OH group is attached.

But this is where Slim and organic chemistry part company. An organic molecule cannot be named for just one characteristic; the molecular name must describe all the features of the molecule so that its name will be unique. To decide what a particular molecule should be called, the name of the basic structure is determined first, usually a carbon-chain backbone such as methyl or ethyl. Then the names of all the functional groups, all the characteristic, identifying groups, are tacked on. As such, the names for organic molecules can become quite ponderous.

Tetramethyldiaminobenzhydrylphosphinous is said to be one of the longest dictionary entries, but the length is not all that unusual for organic chemicals. Some chemists have been known to wax creative when presented with the opportunity to name a new base structure. One new material was dubbed buckminsterfullerene because it resembles a geodesic dome, a design originated by the architect R. Buckminster Fuller. Barbituric acid, the starting material for the family of compounds known as barbiturates, is said to have been named after Barbara, the love interest of the scientist who first isolated the compound.

Aside from finding architects and lovers to name compounds after, organic chemists also spend their time investigating the amazing number of reactions that each one of these compounds can undergo. As we saw in our demonstration, moving or removing just one part can radically change the qualities of the material. So the shelves of chemistry libraries bulge with journals devoted to the intricacies of organic chemistry. And a great deal of work remains to be done. As our feedstock shifts from petroleum to the biomass or the reuse of recycled materials—as it inevitably must—organic chemists are revisiting classic chemistry and inventing chemistry that is new. But remember, all that organic goo must connect with a rocky, sandy soil; an ocean of water; and an atmosphere of air. The inorganic world is where we venture next.

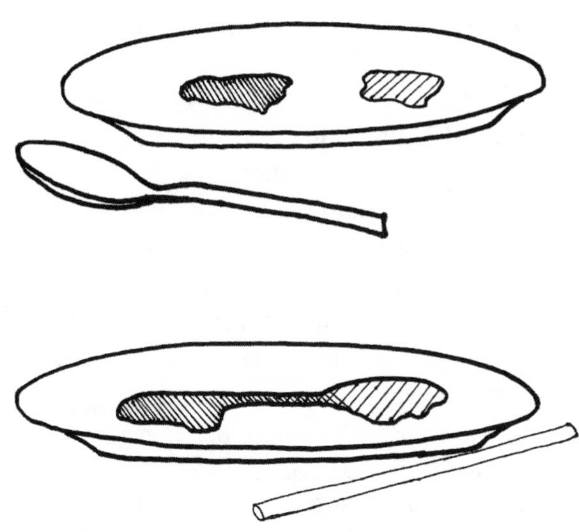

# Inorganic Chemistry Demonstration: Entropy, Ammonia, and Amoeba

*Another year's installment of flowers, leaves, nightingales, thrushes, finches, and such ephemeral creatures, took up their positions where only a year ago others had stood in their place when these were nothing more than germs and inorganic particles.*

—Thomas Hardy, *Tess of the d'Urbervilles*, ca. 1890

Take a glass of water and some of the food-coloring drops that were suggested for purchase in the "Shopping List and Solutions." Very gently add a drop of food coloring to the glass of water, holding the food-coloring bottle as close to the surface as possible to minimize agitation. Then relax and watch the beautiful ballet as the food coloring diffuses through the water. It will mix spontaneously due to entropy, the natural tendency of systems to go toward a state of maximum disorder.

Now, making certain your safety glasses are in place, take the iron acetate solution mentioned in the "Shopping List and Solutions" and put a puddle of it, about a half inch (about 1 centimeter) in diameter, on a dry ceramic or glass plate. Take care that the puddle does not smear. If the puddle clearly has a raised top, rounded by surface tension, then you are ready for the next step. If not, there is probably some sort of residue on the plate. Get a clean, dry plate and try again.

Take some drops of household ammonia and make another puddle with the same rounded top, positioned about a quarter inch (about a half a centimeter) from the iron acetate puddle. Now carefully draw a line of liquid connecting the two puddles by dragging the end of a straw from one puddle to the other. The puddles should merge just at the point of contact.

Observe the two solutions diffusing into one another. The patterns formed can be quite lovely and delicate, like a liquid garden. The process can be set into slow motion by putting a drop of glycerin or mineral oil between the two reagents and connecting both puddles to the one in the middle. A similar reaction-diffusion pattern can be seen by joining a puddle of copper sulfate solution and ammonia or copper sulfate and baking soda solution.

The mixing is, once again, spontaneous, but there is an added element of entertainment in that a chemical reaction is taking place, too. Watching the amoebaelike progress of the two inorganic solutions toward each other and the waves of reaction spreading fingerlike across the plate, one can imagine the first formation of structures in the primordial soup and the slow, steady progress out of the mire.

# CHAPTER 2
## Chemistry Rocks

*Once you are there you'll be like a drop of water in a piece of rock crystal—your medium will dignify your commonness.*

—Thomas Hardy, *A Pair of Blue Eyes,* ca. 1900

We hope that we've convinced the reader of the significance of organic chemistry, but not to the detriment of inorganic chemistry. From gemstones to limestone and nitrates to phosphates, inorganic chemistry is of enormous importance, too. If organic chemistry can be described as the chemistry of covalent carbon, then inorganic chemistry can be described as the chemistry of everything else—and everything else covers quite a lot. In fact, many of the materials we have already discussed are inorganic: metals, semiconductors, salts, and many acids and bases. Yet our cultural emphasis on organic compounds, stemming from the abundance and availability of organic starting materials in the form of petroleum and coal, has led us into an organic mind set. And this single-mindedness may have resulted in an unfortunate dependence on a limited resource. However, a look back at the elemental abundance found on

Earth's crust tells us that oxygen and hydrogen are much more abundant than carbon (oxygen would be the Godzilla on the graph and carbon would be the mouse), so there may be alternatives. Fuel cells that use hydrogen and oxygen are currently powering the space shuttle and have the potential to power cars as well as DVD players. It is nice to know that even though it will be difficult to perfect, the future holds alternatives to fossil oil. Here we will take a look at some of the topics that are currently the special purview of inorganic chemistry: coordination complex compounds of transition metals, representative-group chemistry, and radiochemistry.

The transition metals are located in a big block of elements that form the "transition" from the nonmetals to the metals, as shown in figure 2.2.1.

| H | | | | | | | | | | | | | | | | | He |
|---|---|---|---|---|---|---|---|---|---|---|---|---|---|---|---|---|---|
| Li | Be | | | | | | | | | | | B | C | N | O | F | Ne |
| Na | Mg | | | | | Transition Metals | | | | | | Al | Si | P | S | Cl | Ar |
| K | Ca | Sc | Ti | V | Cr | Mn | Fe | Co | Ni | Cu | Zn | Ga | Ge | As | Se | Br | Kr |
| Rb | Sr | Y | Zr | Nb | Mo | Tc | Ru | Rh | Pd | Ag | Cd | In | Sn | Sb | Te | I | Xe |
| Cs | Ba | La* | Hf | Ta | W | Re | Os | Ir | Pt | Au | Hg | Tl | Pb | Bi | Po | At | Rn |
| Fr | Ra | Ac† | Rf | Db | Sg | Bh | Hs | Mt | • | • | • | | | | | | |

| *Ce | Pr | Nd | Pm | Sm | Eu | Gd | Tb | Dy | Ho | Er | Tm | Yb | Lu |
|---|---|---|---|---|---|---|---|---|---|---|---|---|---|
| †Th | Pa | U | Np | Pu | Am | Cm | Bk | Cf | Es | Fm | Md | No | Lr |

Figure 2.2.1. The transition metals are found in the block in the middle of the periodic table.

These elements have a peculiarly colorful chemistry that stems primarily from their facility at assuming different oxidation states. As you may recall from our discussion of redox chemistry, most elements are fairly happy to share electrons with other elements so that they all wind up with the optimum number of electrons. The nice thing about transition metals is that most can be satisfied with two or more different electron configurations. Recall from the redox demonstration that light orange iron acetate turned green with the addition of ammonia and then red with the addition of hydrogen peroxide. The hydrogen peroxide caused iron to shift from an ion with two electrons missing ($Fe^{2+}$) to an ion with three electrons missing ($Fe^{3+}$). The first color change (orange to green) was the result of ammonia crowding around the iron. The second color change (green to red) was a result of the ammonia molecules interacting differently with the new ion. How closely surrounding ions or molecules are held to a charged ion and the way they arrange themselves around the charged ion is called their *coordination* with the ion. The resulting structure, ion and coordinating groups, is called the *coordination complex*. The study of the coordination complexes formed between transition metal ions and the molecules that surround them is called *coordination chemistry*.

The study of coordination chemistry stems from the observation that nothing is ever completely alone in solution. Even the most aloof ions, in solution, are surrounded by some entourage of counter-ions (ions with the opposite charge) and/or solvent. Because transition metals nearly always form positive ions (the rare exceptions occur with exotic metals or under exotic conditions), they are surrounded by water molecules, at the very least, with the more negative oxygen end of water closer to the positive ion. If there are other ions or molecules available in solution, such as the acetate ion or ammonia as in our demonstration, these too can coordinate with the central metallic ion, as shown in figure 2.2.2. There is generally room for four or six of these companion species, called *ligands*.

A ligand is any small molecule or ion that is attracted to and surrounds a metal ion. Common ligands are $H_2O$, $NO_2$, $Cl^-$, and $NH_3$ (water, nitrogen dioxide, chloride, and ammonia). Depending on the nature of the ligand and the charge on the ion, some are held relatively tightly and others are held loosely.

As we noted in our discussion of photochemistry—and as we encounter every day in our use of items from whitening toothpaste to rose-colored glasses—light interacts with chemicals. Whether they are brighteners or dimmers, these interactions occur on the molecular level. For inorganic coordination complexes, the color of light that interacts with the molecule will depend on the nature of ligands and how tightly or loosely they are held to the metal ion. Due to a complex combination of influences, including electrical attraction, intermolecular forces, and

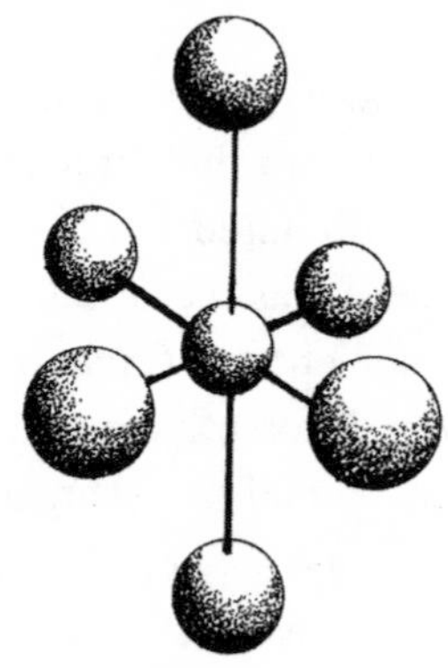

Figure 2.2.2. A central metal ion surrounded by six companion molecules or ions.

entropy, ammonia is strongly attracted to the iron ion. When ammonia was added to the iron solution, the ammonia molecules crowded toward the metal ion, displacing some of the water molecules and acetate ions, and in the process changed the color of the compound. When we added the peroxide, we changed the oxidation state of iron and how tightly the ligands were held. This rearrangement changed how light interacted with the complex—in other words, it changed the color.

Larger organic molecules can form complexes with metal ions, too. Radiator cleaner consists of a water-soluble complexing agent, such as the organic compound oxalic acid, which forms a water-soluble cage around metal ions and lifts them from the walls of the radiator. The metals found deposited on radiator walls can include iron, in the form of rust, and calcium. We don't normally think of calcium as being a metal, but it is. Calcium is located on the left side of the periodic table, which makes it a metal. Iron can form coordination complexes like other transition metal ions. When iron forms a complex with oxalic acid, the iron becomes water soluble.

Organic complexing molecules that move metal ions into solution are known at *chelating agents.* Perhaps the most important organic chelation is that which occurs in our own bodies. Most of the trace metal ions needed for life are bound to protein molecules in some fashion or another, and the protein can be thought of as chelating agents. Hemoglobin is, in one respect, a chelated form of iron. Without the hemoglobin-iron coordination complex, oxygen could not be transported through our blood. As the protein in hemoglobin clusters itself around the iron ion, it creates a neat pocket for the oxygen gas. Blood-borne hemoglobin then carries this oxygen to where it is needed by the cells.

The block of two columns on the extreme left of the periodic table (including calcium) and the six columns on the extreme right of the periodic table, as shown in figure 2.2.3, are the *representative* elements.

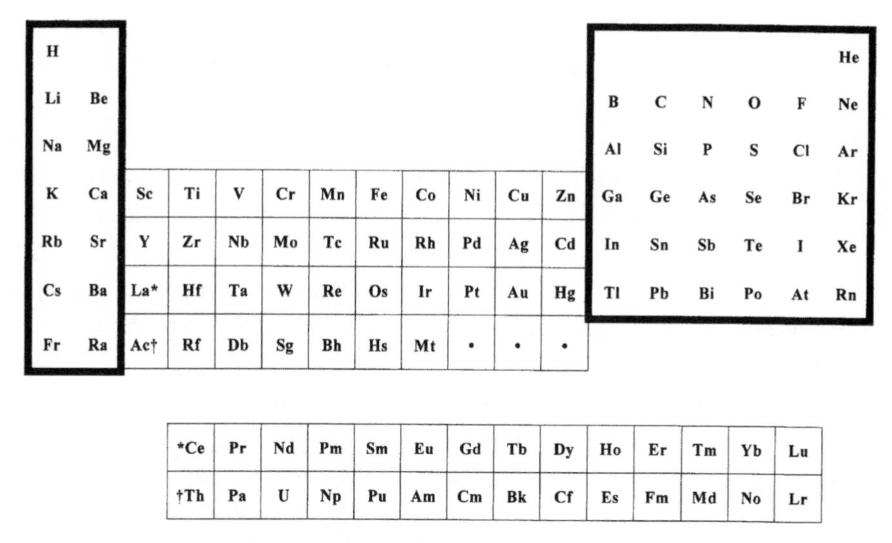

Figure 2.2.3. The representative elements.

These elements are a diverse bunch, containing as they do metals and nonmetals and, at room temperature and normal pressure, some gases, some liquids, and some solids. Within the representative group are several important families of elements with chemistries all their own.

The first column on the far left contains the family of alkali metals, several of which are familiar: lithium (Li), sodium (Na), potassium (K). Most of these elements, as well as the rest of the representative metals, are never found in a natural metallic state. They are present in coordination complexes or in salts. In tales of the Death Valley desert, a generic character named Alkali Sam or Alkali Pete or just old Alkali often crops up, a name derived from the desert salts surrounding this salty character.

Sodium and potassium salts are found everywhere in minerals and soils, and, consequently, natural bodies of water. Their presence in ocean waters had led to their inclusion in the organisms that evolved there. Nowadays a health concern is associated with large concentrations of sodium ions in the body, but the emphasis here should be put on *concentration.* It is not so much the presence of sodium that causes a problem as it is the absence of water. As a culture we have turned to added sodium and away from water, which can be a debilitating combination. It is a sad denial of our saltwater beginnings to tip the balance this way.

The second column from the left contains the alkaline earth metals, beryllium, magnesium, calcium, strontium, barium, and radium (Be, Mg, Ca, Sr, Ba, and Ra, respectively). Magnesium and calcium are present everywhere and are needed by our salty bodies and the salty bodies of our fellow creatures. Calcium is vital to bones, teeth, seashells, and exoskeletons. Calcium plays a critical role in the operation of our muscles as well as communication between cells. Because strontium is in this family, radioactive strontium, a fission product of certain atomic reactions, can be absorbed by the body and used as it would use calcium. Radium, another radioactive element, is also found in this family.

The third and fourth columns of the representative group are over on the right-hand side of the periodic table, a grouping that could be described as the "semiconductor" metals. They make good semiconductors because of their unique location on the periodic table. They are close enough to the transition metals to be somewhat conductive like the transition metals, but they can form covalent bonds like the other materials on the right-hand side of the periodic table. Combinations of tin, lead, gallium, indium, and aluminum (Sn, Pb, Ga, In, and Al, respectively) have been used to make semiconducting materials that have a wide range of properties. Gallium is unique in that it is a solid metal that will melt at body temperature, so it will turn into a puddle when held in the hand. Aluminum, as we have seen, is one of the most common elements in Earth's crust; however, it was not widely used as a metallic material until the early 1900s because aluminum is very reactive. Before aluminum could be fully exploited, a method had to be found for making the metal from the salt. The importance of recycling aluminum metal does not arise from its rarity but rather from the amount of energy and materials that must be expended to claim it from the rocks and purify it.

In the representative elements we find the block that contains carbon, nitrogen, oxygen, phosphorus, and sulfur—the elements, along with hydrogen, that form the basis for life. But just one chapter back we called the chemistry of hydrocarbon derivatives "organic chemistry" and stated we would treat organic chemistry and inorganic chemistry as separate topics. Are we now allowed to blur the lines and include the elements of organic chemistry with inorganic chemistry? Not only are we allowed to, but we must. It was not the chemists who blurred the lines but nature itself. In our interdependent, symbiotic world, there can be no complete distinction between the workings of animal and earth, though for the convenience of the chemist we may seek to impose one. Nowhere does this interdependence show up more impressively than in the inorganic/organic cycles with which elements such as carbon and nitrogen are alternatively used by life, returned to the earth, then the atmosphere, and then reclaimed to further the cause of life.

In the nitrogen cycle, atmospheric, inorganic nitrogen, $N_2$, reacts with atmospheric, inorganic oxygen, $O_2$, in a reaction sparked by lightning. The nitrogen oxides thus formed, NO, $NO_2$, $N_2O_4$, can dissolve in rainwater, which brings them within reach of certain soil bacteria. These bacteria can use them to manufacture the nitrogen-carbon-oxygen-hydrogen organic molecules known as amino acids. Plants and then animals take in the nitrogen-containing organic compounds created by the bacteria, metabolize them, excrete them, and return them to the cycle as organic and inorganic compounds and salts. In Edgar Allan Poe's short story "Cask of Amontillado," he used the presence of nitrate salts to indicate death and decay.[1]

> "The nitre!" I said; "see, it increases. It hangs like moss upon the vaults. We are below the river's bed. The drops of moisture trickle among the bones."

To complete the cycle, another bacterial bug that flourishes under anaerobic (i.e., oxygenless) conditions feeds on nitrogen-containing compounds of decay and returns nitrogen gas, $N_2$, to the air.

Similarly, carbon cycles through an inorganic and organic existence. Carbon (C) begins as inorganic carbon dioxide in the air and is fixed through photosynthesis into organic sugars in plants. These sugars—metabolized and excreted or exhaled by animals, plants, and bacteria—return as carbon dioxide to make another round.

A companion element to carbon, directly below it in the periodic table, is silicon. Silicon (Si) can also form long chains of linked silicon nuclei, and it also assumes a tetrahedral shape like carbon. But silicon does not readily form the double and triple bonds of carbon, so it does not appear to possess the versatility necessary for a complex life form, though some science fiction writers would have it otherwise. Still, silicon forms the basis for quartz, gemstones, and glass, which contribute to the beauty of life.

The rest of the representative-group elements, to the right on the periodic table, are again better addressed as families, this time single-column families. The elements in the column that includes fluorine (F), chlorine (Cl), bromine (Br), and iodine (I) are called the *halogen* family, as in a halogen lamp. In a halogen lamp, there is a gas-phase halogen gas in a small bulb within a bulb. (You can see it if you look closely.) When the tungsten filament of the lamp heats up, a small amount of its surface vaporizes. The purpose of the halogen gas is to keep the filament from vaporizing completely. When the tungsten atoms fly from the surface, they encounter the gas and are brought back toward the surface. This redeposition considerably increases the filament lifetime. Halogens are also found extensively in nature, especially in the ocean and in minerals and, once again, in living things. Iodine is an essential element in the human body. It is necessary for the functioning of the thyroid gland, which is why salt manufacturers add iodine to salt.

To the extreme right on the periodic table are the noble gases—helium, neon, argon, krypton, xenon, and radon. These, as pointed out previously, are very unreactive, which is why we give children helium balloons without worrying about explosions. (We don't give them to babies, but this is because of the balloon material, not the helium.) Radon is notorious as a source of radiation exposure precisely because it is a noble gas—unreactive and a gas—and can percolate through soil and permeate homes that have poor ventilation. Though chemically unreactive, radon can undergo radioactive decay. If inhaled, the emitted particles can damage lung tissue.

Radon is located toward the bottom of the periodic table, as are almost all of the radioactive elements. Unfortunately, the word *radioactivity* conjures up images of bombs, disease, disaster, and destruction. On an intellectual level, it can be acknowledged that radioactivity has provided us with x-rays, imaging techniques, and effective, noninvasive treatments of disease. But on a gut level, we have been conditioned to react negatively to all things associated with radioactivity. However, danger from chemicals is always relative—too much aspirin can be as deadly as too much water if you can't swim—and such is the case with radioactive chemicals. While radioactivity and the elements that produce it should be treated with respect, if handled properly they do not present an uncontrollable danger.

Radiochemistry, the chemistry of radioactive elements, is essentially dominated by the same type of coordination chemistry as the transition metals. One of the worries we have concerning nuclear proliferation is that the chemistry of the uranium and plutonium is really rather straightforward inorganic chemistry. Most of the radioactive elements are found in the group called the *rare earth elements* (those last two detached rows at the bottom of the periodic table), though there are elements with radioactive isotopes scattered throughout the periodic table. As noted earlier, isotopes are atoms of an element that differ only in the number of neutrons: the same number of protons, but a different number of neutrons. Some isotopes possess an unstable number of protons and neutrons. These radioactive isotopes fly apart in an effort to achieve stability, releasing particles and a great deal of energy as they do.

Carbon has a common radioactive isotope, as does potassium, so anything that is living or was ever living or contains something that was once living has some amount of radioactivity. The presence of radioactive carbon is used for *carbon dating*. The radioactive isotope that makes carbon dating possible is called carbon-14 because the number of protons and neutrons in the nucleus totals fourteen (six protons because it is carbon and eight neutrons for this isotope of carbon). Carbon-14 is continuously being produced by cosmic rays bombarding the upper atmosphere. Taken in as carbon dioxide by plants and consumed as sugar and starch by animals that eat plants, carbon-14 becomes uniformly distributed throughout living things and remains at a remarkably constant level. That is, as long as the living things are still living. As soon as a living thing ceases to live, it no longer takes in carbon-14 and the carbon-14 in the dead thing begins to revert to carbon-12. By comparing the amount of carbon-14 in a dead material with the amount that would be expected in living material, scientists can judge approximately how long something has been dead. This technique is limited to dating plant- or animal-derived artifacts less than fifty thousand years old, but that's still pretty good. The technique of carbon dating was used to determine that the Dead Sea Scrolls are about two thousand years old.[2]

We also encounter radioactive elements every day in devices that normally perch quietly and inconspicuously on the ceiling of kitchens and halls: smoke detectors. Smoke detectors? Radioactive? Yes. The working element of a smoke detector is a radioactive source, a sample of americium

(Am), which sends a stream of radiation to a detector. The smoke alarm goes off when this stream is interrupted by smoke—or steam, which is why smoke detectors in the kitchen can have an annoying habit of going off even when there isn't a fire. But smoke detectors have saved an impressive number of lives, so resist the temptation to disable the annoying smoke detector and provide the kitchen with better ventilation instead.

Radiation, from radioactive elements, has found another application that is perhaps a bit more controversial than smoke detectors—the use of radiation to kill bacteria on food. The method does work well; it extends the shelf life for many foods. But some people object to using food that has been irradiated, perhaps for fear of contamination.

Contamination should not be a concern. Food that is irradiated does not come in contact with the radioactive source. There are, however, other thought-provoking objections. Some are concerned that the radiation may initiate reactions in the food itself and could form compounds not normally present. Others say we should worry about anything that will kill virtually all of the bacteria present on an item of food. Why? Because bacteria are not necessarily a bad thing. Bacteria are part of the grand scheme. Nature is very good at preserving life, including the life of bacteria, and has no particular bias for human life. So if we manage to kill off some bacteria, that would create a void, and a new kind of bacteria might evolve to fill that void. We wouldn't be able to predict if they'll be human friendly or not. This adjustment on the part of nature is possible because natural systems are fluid—that is, they change. We tend to think of our world in terms of permanence because, for the most part, the adjustments are very slow. But biological systems respond to stress, as do the chemical reactions from which they are formed. Further discussion of biochemical molecules is the topic we take up next.

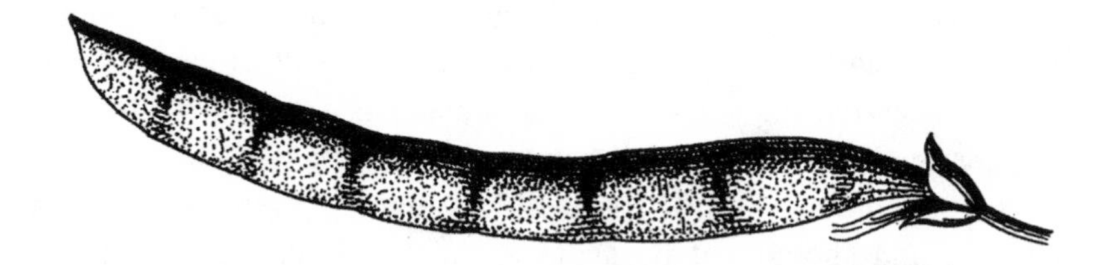

# BIOCHEMISTRY DEMONSTRATION: FAT, FLATULENCE, AND BEAN SOUP

*How on earth are you ever going to explain in terms of chemistry and physics so important a biological phenomenon as first love?*

—Albert Einstein, ca. 1940

It is a fact of life: we need to eat healthy foods, and healthy foods give us gas. The problem is that the starches in grassy foods are not completely digested in the human small intestine before they enter the large intestine. This inefficiency is no problem for nature, however. In one of those wonderful symbiotic schemes of which nature is so fond, colonies of bacteria, acknowledged delicately as "intestinal fauna," live in our intestine. As we mentioned in our discussion of the properties of the gas phase, these bacteria have the necessary enzymes to digest grassy foods and feed on the leftovers quite happily. In the process, however, they produce gas. The digestive process also produces sulfides, sulfur-containing organic chemicals that we associate with flatulence: that rotten-egg odor produced by bacteria performing the same function in rotting eggs. The odor of flatulence comes from these sulfides, carried with the gas, which is why intestinal gas is sometimes more odiferous than other times.

The problem of odor is an artificial problem, imposed only by societal norms, but the bloating caused by an overabundance of intestinal gas can be distracting at best and debilitating at worst, so there are reasons to avoid an excessive amount. One way, of course, is not to eat the foods that cause gas, such as raw vegetables and beans. But many of these foods are convenient and concentrated sources of essential proteins, vitamins, and roughage, and should not be avoided entirely. If desired, an enzyme, alpha-galactosidase, can be taken before eating and aid in the digestion of starch before the bacteria have a chance to feed. In this demonstration, we will show these enzymes at work.

Donning your safety glasses, make up two solutions of just less than a quarter of a teaspoon of cornstarch (1 milliliter) in one cup (240 milliliters) of water. Crush a tablet of a digestive aid that is meant to be taken before meals to reduce intestinal gas. These pills should contain the enzyme alpha-galactosidase, which will digest starch. Check the label to make sure. Add the crushed tablet to one of the cups of cornstarch solution. Let the glasses sit for about an hour and then add a drop of iodine tincture iodine to each. Iodine is a well-known indicator for starch because iodine forms a lovely blue-colored complex with starch. The solution without the enzyme will turn a violet-blue color, indicating the presence of starch. The solution to which the enzyme was added should remain the brown color of the iodine tincture. If there is a blue color, it will be much weaker. This demonstrates that the starch has been broken down.

Enzymes and starches are under the purview of biochemistry, the kingdom we enter next.

# CHAPTER 3

## The Body of Chemistry Meets the Chemistry of the Body

*For nitrates are not the land, nor phosphates; and the length of fiber in the cotton is not the land. Carbon is not a man, nor salt nor water nor calcium. He is all these, but he is much more, much more.*

—John Steinbeck, *Grapes of Wrath*, ca. 1940

*"Ben, I have just one word for you, one word . . ."*
*"What's that?"*
*"Plastics."*

—Calder Willingham and Buck Henry, *The Graduate*, screenplay based on the novel by Charles Webb, ca. 1960

What do plastics have to do with people? As it turns out, quite a lot. Plastics are made by *polymerization*, the process by which simple molecules are linked together into extensive chains thousands of molecules long. There is not a more proficient polymerizer than a living cell.

From skin to hair and tonsils to toenails, the material that makes us is continuous, pliable, and polymeric. As we pointed out in our discussion of organic chemistry, the success of organic life forms is based on carbon's ability to form long chains, and when units of these chains are repeated, there you have it: polymers. Our bodies, as well as the bodies of everybody and everything else, from amoebas on up, are composed of long chains of repeating fundamental units—*carbohydrates*, *amino acids*, and *nucleotides*—the polymers that are us.

Let's begin with the carbohydrates and the polymers they form. What are carbohydrates? We know that they are something to be avoided by adherents of certain dietary disciplines, but what are they? Originally carbohydrates were thought to be carbon surrounded by water (hence carbo-*hydrate*), somewhat as our inorganic coordination complexes consisted of a metal surrounded by ligands. But we now know that the basic unit of a carbohydrate is a saccharide, which is a ring-shaped carbon molecule, such as shown in figure 2.3.1.

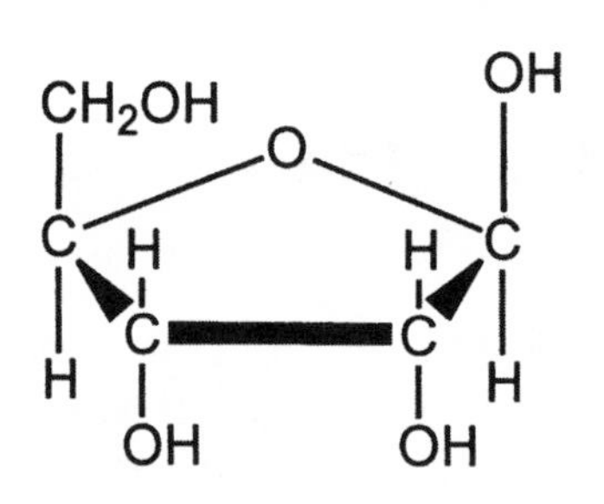


Figure 2.3.1. A saccharide, the basic unit of carbohydrates.

Carbohydrates are sugars—glucose, fructose, galactose. The familiar *sucrose* is a *disaccharide*: two saccharides linked together. When more than two saccharides are linked together, the result is a polysaccharide. Long-chain polysaccharides form the essential materials of starch, cellulose, and glycogen.

As is well known by those who are diet and health conscious, the body metabolizes sugars for energy and, when excess sugar is available, stores the sugar for later use. Animals store sugars as polymers of glucose called *glycogen*. Plants make a slightly different polymer of sugar for storage purposes, and this is what we call starch. Plants also use long chains of carbohydrates to make cellulose, a material for structural purposes, not food-storage purposes. The human body is not capable of recovering saccharides from cellulose, and sometimes indigestible carbohydrates are incorporated in foods to lower the calorie count per item.

Some types of lipids, or fats, are also stored by the body for eventual metabolism, and fats are another food that has been singled out for avoidance because of the negative effects of overconsumption. But there can be no doubt that the body requires fats, especially the growing bodies of children. For instance, vitamins A, K, and D need fat to be absorbed. So the limitation on fats really needs to be tempered by knowledge of which fats to limit and which fats are essential. There are several different types of fats, and these have varied functions.

Two main varieties of lipids are fatty acids and steroids. The basic structures of these compounds appear quite different, as seen in figure 2.3.2, so it may seem curious that chemists lump them together.

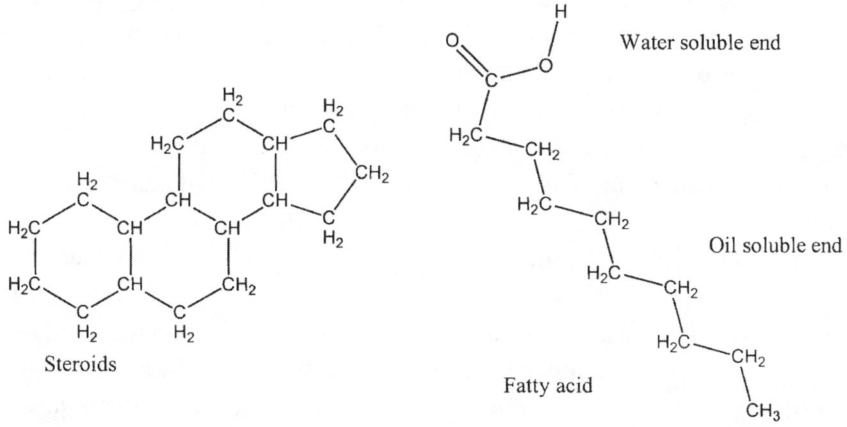

Steroids

Water soluble end

Oil soluble end

Fatty acid

Figure 2.3.2. The two main categories of lipids: steroids and fatty acids.

These compounds are grouped together because they have the common property of being soluble in oily liquids. In the case of fatty acids, one end is soluble in oil and the other end is soluble in water, an ambiguity that turns out to be tremendously useful, as we shall see.

The fatty acids are long chains of carbons terminated by an organic acid group, COOH. In their various configurations, these fats store energy (and are the basic component of all those troublesome fat cells) and also serve to protect and insulate internal organs. They form the insulation on nerve fibers and are the basic component of the cell membrane. One of these long-chain fatty acids, arachidonic acid, can be released when a cell is injured by cutting or smashing. This acid is converted into a compound that stimulates nerve cells (pain), rushes blood to the site (inflammation), and raises the temperature to speed up bodily repair processes (fever). That all this happens is remarkable enough. But the next time you cut yourself, note how quickly it happens. Quickly, but not instantaneously. It takes time for the chemical reactions to take place, but an amazingly short time at that.

Steroids are lipids that have the same basic cyclic structure shown in figure 2.3.2. The steroids have a notoriety of their own, but again, they serve their purpose and cannot be dispensed with entirely. Steroids are components of cell membranes, nerve cell insulation, and the basis for steroid hormones.

Cholesterol is a steroid. And, as everyone knows, eating too much cholesterol can be bad, but there is no way that cholesterol can be entirely eliminated from the diet. The body manufactures cholesterol because the body needs cholesterol for many essential functions. So what is the problem with cholesterol? The problem is that when there is too much in the body, it can accumulate on the blood vessel walls and cause constrictions where the blood should freely flow. Recall that in our discussion of organic chemicals, we said that saturated compounds are ones that have all single bonds, C–C, which means no double, C=C, or triple, C≡C carbon/carbon bonds. It turns out that saturated fats are also the solid fats like lard or solid vegetable shortening. Saturated fats last longer without spoiling, so manufacturers of margarine used to purposefully *hydrogenate* their products—that is, break up double and triple bonds by adding hydrogen—to prevent spoilage. However, it is now recognized that saturated fats can stimulate the body to synthesize cholesterol and inhibit its elimination. Unsaturated fats, on the other hand, seem to promote the elimination of cholesterol.[1] The quantity of cholesterol, however, is not the whole problem.

Unwanted deposits of cholesterol occur because cholesterol, being a lipid with no water-soluble end, needs help to be transported around in the bloodstream. The transporters are called low-density lipoproteins (LDL). They form little bubbles around the cholesterol, not unlike soap solvates dirt, using their oily ends to connect with the cholesterol and their organic-acid ends to connect with the watery fluid of blood. The structures of the LDL help them deposit their cholesterol load when they reach the cells where it is needed, but unfortunately, their structure also facilitates the dumping of excess cholesterol on blood vessel walls when the cells have all they need. When we speak of "bad cholesterol," what we really mean is the LDL-cholesterol complex. People who are prone to high blood cholesterol levels should avoid eating cholesterol, but they also must avoid saturated fats because these fats contribute to LDL production.

There is, however, another structure, called high-density lipoprotein (HDL), that will wash the excess cholesterol from the walls and down to the liver to be broken down. When we speak of "good cholesterol," what we mean is the HDL-cholesterol complex. So it might first seem that cholesterol deposits could be avoided by maintaining a proper balance of LDLs versus HDLs, but it is more complicated than that. Overall levels of fat must be lowered, too. Excess fats will stimulate the production of cholesterol by the liver, and a person on a high-fat diet can actually reabsorb some of the cholesterol that has been moved to the liver. So there is no free lunch on a high-fat diet.[2] The key to a healthy diet is what we've known all along: balance. Some fruits, some vegetables, some carbs, some fat, some protein.

What are proteins? Polymers, of course—polymers of basic units called amino acids. Amino acids are small organic molecules that contain a nitrogen group and an organic acid. A generic amino acid is shown in figure 2.3.3. The organic acid group is the carbon with the oxygen double bonded and an OH group attached by a single bond. The nitrogen group, $NH_2$, is called the amino group (hence *amino acid*). The "R" to the side is a symbol meant to stand in for various other groups that can be added at this position and in this way make up a variety of amino acids.

Note that there is a carbon with four attachments. Amino acids can have a "handedness," and naturally occurring amino acids have a preferred handedness. If a molecule of the wrong configuration is introduced—if just two bonds are interchanged—the molecule no longer fits and the results can be disastrous.

There are twenty different amino acids that are found naturally in the human body. Polymers of these amino acids form protein structures such as fingernails and cartilage as well as enzymes and hormones. Some of the amino acids can be manufactured in the body, but about an equal number

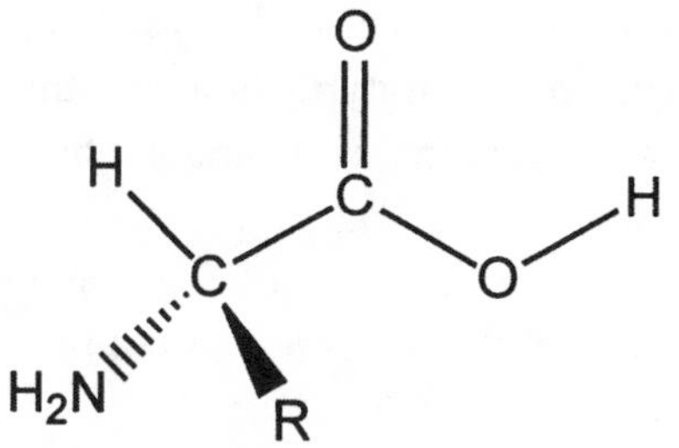

Figure 2.3.3. A generic amino acid.

must come from the diet. All of the necessary amino acids are found in animal protein such as meat and eggs, but they can be acquired from nonmeat sources, too. There is no vegetable that contains all the needed proteins, but as long as many different varieties—squash, corn, rice, beans, soy, and nuts—are consumed, the body's protein requirements can be met.

Of the many materials we have discussed so far, it may be the proteins that perform the most functions. Their long chains include both acid (COOH) and base ($NH_2$), and intermolecular forces allow the protein chains to fold in on themselves in many complex but reproducible configurations. The forces that hold the proteins in their folded form are the same intermolecular forces that we encountered before and as such can be easily disrupted. The vast majority of biologically active proteins perform well only in their natural environs and when disturbed, or *denatured*, by even small changes, can no longer function. The protein of a raw egg can be transformed from its functioning configuration by the addition of even small amounts of heat, which is what cooking is all about. Hemoglobin is a protein that incorporates iron ions. The brown color of cooked or spoiled meat results from a change in oxidation state of the iron in hemoglobin, just as we saw when we changed brown iron acetate to a red complex by adding hydrogen peroxide.

It would not be possible to catalog in this small amount of space the number of materials that proteins form in the body and their many roles, but perhaps another look at one class in particular—enzymes—will serve to communicate the impressive dexterity of proteins.

Enzymes are the catalysts of the body. They facilitate the rapid reactions that are necessary for life. (Recall the message that must be communicated to your brain when you cut your finger, if the communication were not rapid enough, irreparable damage could be done.) Enzymes perform their catalytic function through their three-dimensional structure. Each enzyme has a unique shape that allows it to bond to target molecules. We illustrated one way this binding might speed chemical reactions in our discussion of reaction rates. When two reactants bind to the enzyme, the enzyme can hold them together in optimum alignment until they react.

The manner in which the body uses enzymes is quite elegant.

Enzymes need not always be turned on, continuously catalyzing reactions. If the product of the reaction is not required, the enzyme needs to be shut down. Many times, in a gracefully choreographed equilibrium ballet, the same product that the enzyme produces can work to switch it off. The overabundance of the product shifts the reaction away from producing products to breaking down the product instead. The body must maintain a delicate balance of materials, neither a dearth nor a deluge will do, so the shutdown mechanisms must work as well as the stimulating mechanisms, and amazingly enough, they do.

The number of proteins themselves must also be regulated. The system responsible for protein supply operates on the principle of production on demand. For instance, the enzyme needed to break down lactose would only serve to gum up the works if it were always present in the intestine, just waiting for a dairy product to be consumed. So the body breaks down and recycles the proteins of this enzyme, storing the amino acids in the soup that exists in the interior of every cell. When lactose is again in the intestine, the cells are stimulated to produce more enzyme to deal with it. How does the cell produce more enzymes? With another polymer, of course.

The polymer we are talking about this time, however, is unique in that it is built from three different types of fundamental units: a nitrogen-containing base, a sugar, and a phosphate group, as shown in figure 2.3.4. The linkage is between the phosphate group and the sugar, and, linked together, as shown in figure 2.3.5, they form an intricate scaffolding that twists and turns into the compact material called deoxyribonucleic acid: DNA. The stuff of genes and chromosomes.

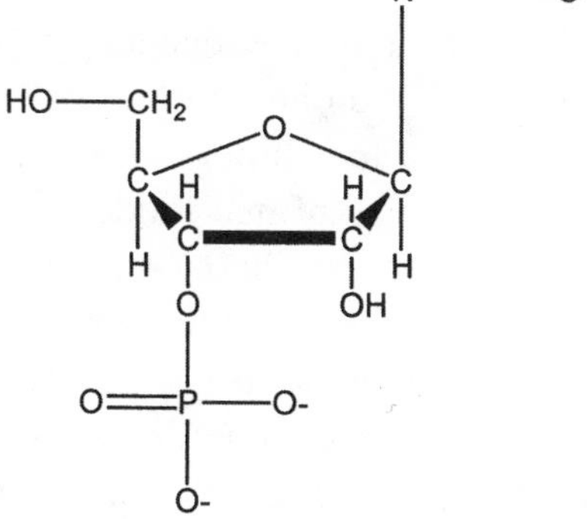

Figure 2.3.4. A phosphate group, a sugar, and a nitrogen-containing base.

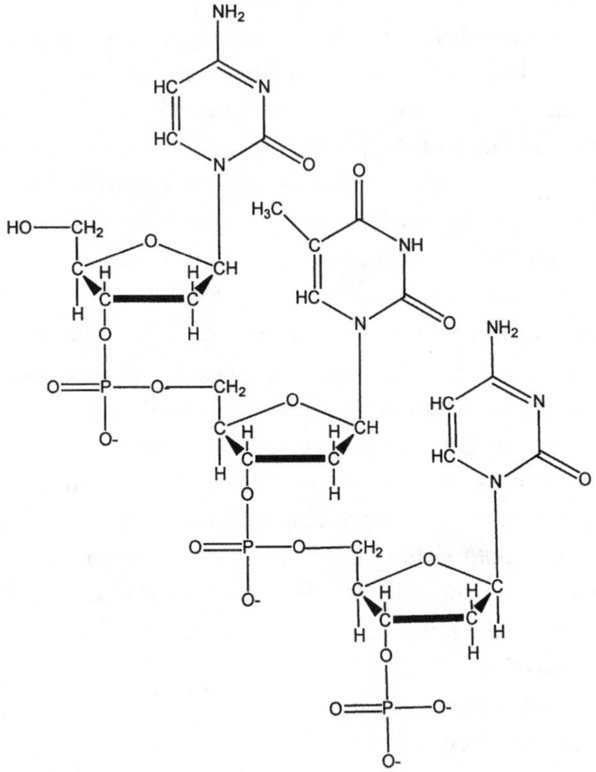

Figure 2.3.5. A segment of DNA showing the linking that occurs between nucleotides.

A chromosome, one long strand of DNA, is a molecule that contains information on how to make all the proteins of the body. A gene is one segment of the chromosome that tells how to make one specific protein. The sequence of nucleotides in a gene serves as a template that is ultimately translated into a sequence of amino acids in a protein.

DNA can be thought of as a diminutive recipe file. When the body needs a hormone, it goes to the DNA to get the instructions on how to make one. When it needs a skin cell or a fingernail, it goes to the DNA to get the recipe for the proteins. When a particular protein is needed—when there is a deficit or an increased demand by the body—stress is put on an equilibrium that causes a section of DNA to be unraveled and a protein to be assembled from its pattern. In this way, when the body needs more hair, the cells are stimulated to produce hair protein. When the body needs more digestive enzymes, the message is received and more are produced. When the body needs adrenalin, the hormone is ready—at a rate that allows us to leap clear of cars or outrun a roach. But, as pointed out in the demonstration with which we began this chapter, for all the amazing ability of the cell to respond to the demands of the body for enzymes, there remains one enzyme that the human body cannot produce: the enzyme to completely break down beans.

Enzymes, such as the one we used in our demonstration, are governed by the principles of chemical kinetics—one of the many links between the basic principles of chemistry and the intricate chemistry of life. Our rapid and cursory survey of biochemistry here, combined with our previous discussions of biochemical systems, shows that in life all our chemical principles come into play: acid-base reactions, redox reactions, chemical bonding, intermolecular forces, concentration, solids and solubility, kinetics, and even phase transitions and the gaseous phase.

Equilibrium plays an essential part, as does the thermodynamics that governs it, energy and entropy alike. Where there is a need of a compound, the body produces it. When there is an excess, the body stops production—and consumes or excretes the excess. When we need energy, we eat. The living body functions by the supply and demand of chemical equilibrium.

It may seem odd to think of entropy in the context of biochemistry because biological systems seem so organized and ordered, but entropy is as big of a force in biological systems as everywhere else in the universe. Diagrams of cells in encyclopedias can leave the false impression that cells are like row houses in a city or apartments in a complex: all stacked and neat and shaped the same. But a look through a microscope will serve to correct this impression. Cells are really very irregular and messy. Each one seems to have its own shape and habit. The cell membrane is not rigid; it is fluid and constantly shape-shifting about. DNA is not a rigid printout of information but contains "jumping genes" and "junk genes," genes that can move around and genes that seem to serve no other purpose than to add to the confusion.

But there are reasons to appreciate all the messiness. The irregularity means no two natural biological systems will ever be completely alike. We have learned to understand this uniqueness and—with the techniques of analytical chemistry that we will discuss next—exploit it. The tools of analytical chemistry now enable us to analyze and extract information from a flake of skin or a fleck of spit. Sherlock Holmes would be proud.

# Analytical Chemistry Demonstration: The Proof Is in the Prints

*Supposing the entity of the poet to be expressed by the number ten, it is certain that a chemist in analyzing . . . would find it to be composed of one part self-interest to nine parts of self-esteem. . . . Gringoire's nine parts of self-esteem, swollen and inflated by the breath of popular admiration, were in a state of prodigious enlargement, obliterating that almost imperceptible molecule of self-interest.*

—Victor Marie Hugo, *Notre Dame de Paris*, ca. 1920

DNA evidence has proven to be very valuable, accurate, and reliable, but it is not always available. Fingerprints, however, are easy to see and are sought at every crime scene. To show how latent fingerprints might be made visual, put on your safety glasses and try the following demonstration.

Take some aluminum foil and shape a small square into a bowl or pan. Take a plastic food container with about a one-quart capacity. (A large butter tub will do nicely.) Wash it and its lid, then rinse and dry these items. Find a small jar, an empty medicine bottle, a smooth drinking glass, or another transparent item that will fit into the butter tub and wash it first completely, inside and out, handling it with gloves or a towel to avoid unplanned fingerprints. Find a clean pair of tweezers or a pair of pliers, and you have all of the items assembled for your fingerprint-developing chamber.

Get the superglue listed in the "Shopping List and Solutions" and make sure it is the kind that contains cyanoacrylate. The fingerprints will be developed on the small jar or similar item that you chose, so firmly grasp the item with your bare fingers, being careful not to smudge the prints, and set it down inside the butter tub. Jot down on a notepad where you touched the jar for future reference. Now take the small foil pan and place about thirty drops of superglue into it. The quantity of glue used at this point depends on the size of the chamber. A larger chamber will require more glue than a smaller chamber. Put the foil pan of glue into the bottom of the chamber next to the jar. You may need to handle the foil pan with tweezers or pliers here to avoid getting glue on your fingers as you lower the foil pan into the chamber. Carefully snap on the lid to the chamber, making certain that the glue doesn't spill from the foil pan, and set the chamber aside overnight. The glue will evaporate over time, and the glue vapor will react with the oils and compounds left behind on the jar by your fingers.

In the morning, carefully open the chamber and remove the jar by inserting a pencil, pen, or other long probe into the jar opening. Try not to smudge the fingerprint residue, which can now be observed on the jar. Are there prints on other parts of the jar that were not noted the previous day? There were on ours, and we tried to be careful. If you have a steady hand, some black paper, and some clear packing tape, you can lift the print off the jar and onto the tape. Carefully place a small piece of tape over your prints, taking care not to smear them, and then remove the tape with the prints. Because the developed prints are light gray, attaching the tape to the black card makes them more visible. Prints can be lifted off many surfaces, some of them pretty unlikely. For instance, with care, prints can be lifted from the *inside* of surgical gloves.

Put on a surgical glove and grasp a doorknob or other item firmly. Peel off the glove so it is inside out and place it into the developing chamber. Spread it out a little so the vapor can reach all areas. Develop the prints as before. We managed to develop one print on the inside of our glove that we could lift off onto tape.

Blood-splatter evidence is also important in reconstructing crime scenes, and this additional demonstration shows how the presence of blood might be detected. Hemoglobin in the blood of mammals acts as a catalyst for the decomposition of hydrogen peroxide. A piece of raw meat will have enough blood for this demonstration, or the next time you purchase a piece of beef you could save the absorbent tissue under it and use that as your source of blood. You will need two small clear plastic cups and some red food coloring as a control. Put a small amount of red food coloring into one cup along with some water and then mix. Take the raw meat or the tissue from under the beef and squeeze a little of the blood into the other cup. Now pour about a teaspoon (5 milliliters) of hydrogen peroxide into each cup. The food coloring will slowly oxidize from the peroxide and the color will gradually fade. But the blood will cause rapid and vigorous hydrogen peroxide decomposition. The difference in the reactions clearly distinguishes the blood from food coloring.

# CHAPTER 4
## Chemist as Analyst

*I found Sherlock Holmes alone, however, half asleep, with his long, thin form curled up in the recesses of his armchair. A formidable array of bottles and test-tubes, with the pungent . . . smell of hydrochloric acid, told me that he had spent his day in the chemical work which was so dear to him.*

*"Well, have you solved it?" I asked as I entered.*

*"Yes. It was the bisulphate of baryta."*

*"No, no, the mystery!" I cried.*

—Arthur Conan Doyle, *Case of Identity*, ca. 1890

*It is easy for sugar to be sweet and for nitre to be salt.*

—Ralph Waldo Emerson, *Representative Men*, ca. 1850

Chemist as analyst? Certainly. But we don't use the couch to analyze a chemist; we use chemistry to analyze the couch. We have saved our discussion of analytical chemistry for near the end because in many ways it brings out the best in chemists. In analytical chemistry, all the tools of chemistry and all the talents of chemists are brought to bear on two questions: what is it, and how much of it do we have? At some point, all chemists are analytical chemists and all chemistry requires analysis. There are several specialized areas of concern for analytical chemists such as the quality-control experts that ensure the quality of our food, the reliability of our medicines, and that every new car that rolls off the assembly line will have the same luster. But analytical chemistry is so pervasive that many of us routinely act as analytical chemists when we use tap-water-testing kits, pool-testing kits, radon-testing kits, fish-tank-testing kits, soil-test kits, lead-paint-testing kits, and even pregnancy-testing kits, sugar-testing kits, and the infamous Breathalyzer. The job of the analytical chemist is to find—or invent—the tool or tools necessary to determine the quantity of a given material or the nature of unknown materials.

Analytical chemistry is investigative work. Like a detective solving a mystery, an analytical chemist must proceed very carefully and painstakingly to find the correct answers. Just as the first officer at a crime must secure the scene and prevent evidence from being corrupted, analytical chemists must ensure that all the equipment and glassware to be used is scrupulously clean and free of contamination. Like a police investigator carefully collecting the minutest of evidence, analytical chemists sometimes deal with microgram (a millionth), nanogram (a billionth), and picogram (trillionth) quantities to get their answers. To get an accurate weight, the analytical chemist may even need to take into account the buoyancy of air.

Just as police officers need to maintain their weapons and routinely put in time on a shooting range, analytical chemists must always ensure that their instruments are calibrated and functioning properly. In police work, even if the butler confesses, without the proper police procedure, the most damning evidence can be thrown out. In the same manner, analytical chemists must be able to defend their results by reporting parameters such as *confidence limits*, *instrument noise levels*, and *significant figures*.

Analytical chemists must always be thorough. One thread from a carpet will not serve to locate a suspect at a crime scene, but a thread from a carpet, plus a hair from the dog, plus a smear of fresh paint may convince a jury. Many times an analytical chemist, too, must supply a body of evidence: corroborating results using alternate methods and many repetitions of the same measurements to show, statistically, that the result is reasonable.

Like a scientific sleuth, an analytical chemist must sometimes reconstruct an event. When there is an explosion, there may be many causes, not all of which are malicious. But if traces of chemicals are found that have no other use but to make explosives, then that is good evidence the explosion was planned. When drug use is suspected, blood is examined, but evidence of heroin in the blood is not based on finding traces of heroin. It is based on finding heroin's metabolism product, morphine. Even the presence of morphine does not always indicate heroin consumption. Poppy seeds used for baking have been found to contain traces of opium and can produce traces of morphine in the urine. Prescription drugs can be metabolized by the body into what are, in essence, controlled substances, so other evidence must confirm or refute the presence of illicit drugs. In fact, the parallels between a professional analytical chemist and a criminal investigator are so numerous that we are going to use a hypothetical investigation by a forensic analytical chemist to illustrate how this important specialist proceeds.

It is the task of the forensic chemist to apply analytical chemistry to questions of interest to the legal system. The forensic chemist usually works at a government laboratory and may have the responsibility of going to the scene in question, collecting evidence, testing the evidence, and testifying in court. To show how all of this might work, let's begin with a hypothetical crime scene scenario and follow the progress of the forensic chemist as she proceeds.

Let's say that the police are called one night to an apartment building because of loud shouts coming from an apartment followed by the slamming of doors. When the police arrive they find the body of a woman lying in a pool of vomit. The medical examiner states from his preliminary exam that it looks as though the woman died an accidental death from anaphylactic shock, though he must, of course, perform an autopsy to be sure.

The cause of many deaths every year, anaphylactic shock is an allergic reaction so severe that it can cause the throat to constrict to the point of asphyxiation. The substance causing the allergic reaction, the *allergen*, could be nuts, eggs, bee venom, or any of a number of common substances. People prone to anaphylactic shock have usually suffered an initial, milder reaction to the same allergen, and then their sensitivity increases dramatically with each exposure. People aware that they are prone to such extreme reactions carry an *epipen*, a small penlike device for injecting a dose of adrenalin, which will counteract the swelling and could save a life. According to the medical examiner in our story, the woman found in the apartment has every sign of having suffered this unfortunate but natural death. However, there are the reports of the loud shouts and the slamming door, so a forensic team is called in.

A forensic chemist routinely surveys and records all possible evidence at the scene. In this case, the chemist examines the contents of the cupboard and medicine cabinet and takes swabs of surfaces to see if she can determine the source of the allergen. She swabs the victim's hands and face to collect any chemical residue. She takes a sample of the victim's blood and stomach contents, which are available in the pool of vomit. She notes that the food in the cupboards has been carefully chosen to be free of peanut oil or peanut products, so she suspects that the woman had an allergy to peanuts, a common culprit in anaphylactic shock. She carefully samples all the food in the refrigerator and all containers on the shelves. She examines the pockets and purse of the victim but does not find anything worthy of note. She opens drawers in the kitchen and bedroom with the same result. She looks in the medicine cabinet and sees only common over-the-counter pharmaceuticals such as aspirin, antacid, and bandages. As she packs up and gets ready to leave, she stops for a minute and tells the officer in charge that additional investigative work needs to be done. Based on just what she has seen so far, she believes it was murder. Do you?

While the police officers act on her suggestion, our forensic chemist goes back to her lab to confirm some of her suspicions. To begin with, if the cause of death was a severe reaction to peanuts, then there should have been peanut oil or peanut residue somewhere on the victim's person. The chemist takes her swabs and her samples and begins her analysis.

The first problem confronting our forensic chemist may be separation. Police will separate witnesses, taking them into individual rooms to hear their accounts to avoid interference from other principals. Likewise, an analytical chemist will strive to separate the components of an unknown mixture so that the separate materials can be analyzed without interference from other species. If the components are in separate phases, then a straightforward technique such as filtration can be used. For instance, our forensic chemist filters her sample of stomach contents then cools the filtrate to see if more materials will separate out.

If the component of interest is organic, it can be separated from its inorganic companions by a simple extraction such as was used to extract acetylsalicylic acid from aspirin in the organic chemistry demonstration. But the result of an extraction of stomach contents and blood will be a mixture of organic materials. If the attempt is to separate two organic materials, then a method such as *chromatography* might do the trick. Chromatography is a separation technique in which a solvent washes over a mixture and carries the various solutes different distances. An analogy might be found in panning for gold: the water continually running over the sand washes away the less dense particles and leaves behind the gold.

The technique of chromatography can be demonstrated in an uncomplicated fashion using the acetone found in fingernail-polish remover and some tree leaves. Crush and tear apart two or three green, fresh tree leaves and put them in the bottom of a glass drinking glass. After putting on your safety goggles, add about three inches of acetone (fingernail-polish remover), which should just cover the pile of leaves. Let this leaf mixture rest for several minutes so that the acetone can extract compounds from the leaf.

Cut two strips of paper towel long enough to reach from the top to the bottom of the glass. Wrap one end of the paper-towel strips around a pencil and secure them with a piece of adhesive tape. The paper-towel strips should now be just a bit shorter than the height of the glass, just long enough to touch the liquid.

Lay the pencil across the top of the glass and dangle the paper towel strips into the acetone-leaf mixture. Over a period of several hours, you should see at least two distinct bands of color on the paper towel as the acetone moves up the towel by capillary action. The acetone is carrying the leaf color pigment with it, but how easily the pigment travels with the solvent depends on the type of pigment, so when there is more than one pigment present, they separate into bands. Take a moment and look at the colors of the bands. One band should be green, but the other band or bands may be yellow, orange, or red. In the fall, we say that leaves "turn color," but, in fact, the color is always there. It is actually the chlorophyll that is missing in the fall, and with the chlorophyll gone the other colors come through.

The acetone/towel/leaf system is an example of *solution chromatography*, but there are other chromatographies that work on the same principle: different materials can separate when they travel with their solvent along an absorbent surface. In gas chromatography, the solvent is a gas, and the gas blows over a column of solid particles. The solutes picked up by the gas will separate into bands as the material dissolves, is reabsorbed, and dissolves again in the gas. Ion chromatography, which uses water as a solvent, can be used to sort inorganic salts. High-performance liquid chromatography (HPLC) uses high pressure to force liquid samples through a densely packed column and achieve an even greater degree of separation.

Once our forensic chemist has separated all the organic material from her swabs, her next step might be concentration. The problem of concentration has to be addressed if the *analyte*, the substance being tested for, is present in only small amounts. Although analytical methods are improving all the time and the sensitivities of instruments are constantly increasing, there is still always some lower limit below which the analyte cannot be detected by a particular instrument. For instance, it may be difficult to test for small quantities of a toxin in the blood. Our bodies naturally concentrate toxins in the liver, and the analyst may take advantage of this fact if the liver is available, but if it is not, concentration may be achieved by evaporation, precipitation, or centrifuging—a process wherein a test tube sample is spun at a high rate of speed until the heaviest components settle to the bottom.

A different approach is taken with small samples of DNA. With a technique termed the *polymerase chain reaction*, or PCR, the DNA is not concentrated; instead, DNA is multiplied. The polymerase chain reaction uses the body's own chemical machinery, harvested from cells and brought to a test tube, to make copies of DNA until the concentrations are high enough to analyze.

Because the analyte she is after, peanut residue, is organic, the analytical method of choice for our forensic chemist will probably be GCMS, which stands for gas chromatography with mass spectroscopy. In this technique, the material is separated into its components using chromatography, as outlined above. As the carrier gas, analogous to the acetone in the leaf pigment demonstration, flows over the sample, the components of the mixture are separated into bands that progress down a tube with absorbent packing called a *column*. The challenge is to analyze them on the other end of the column as quickly as they exit the column. The analysis can be accomplished by a number of methods, but one of the most popular is mass spectroscopy.

*Spectroscopy* is the general name for any technique that has a *spectrum* as its final result. A spectrum is a graph of two variables: the quality being tracked and its quantity. For instance, people in the United States display a wide spectrum of height, ranging from very short to very tall. If we were to make a plot of the number of people versus relative height, we might have a spectrum such as shown in figure 2.4.1.

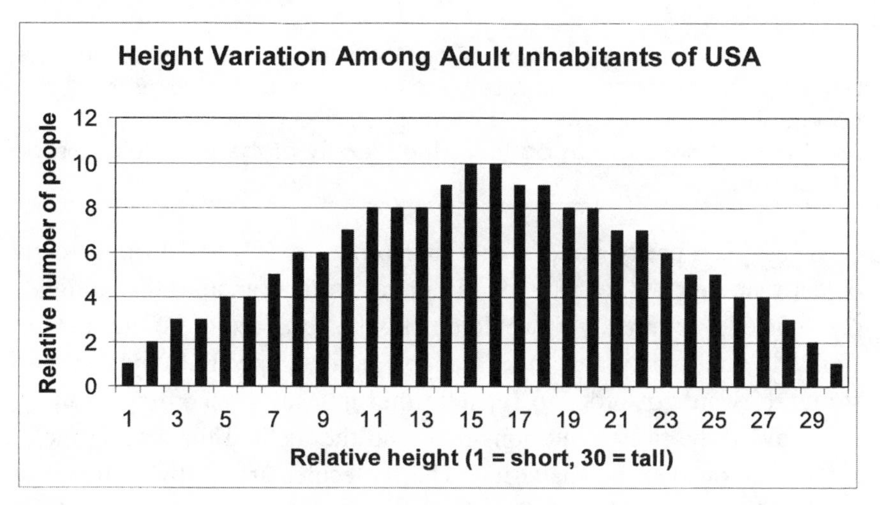


Figure 2.4.1. A hypothetical spectrum of height for people in the United States.

In mass spectroscopy, the fragments of molecules are sorted by mass, as shown in figure 2.4.2.

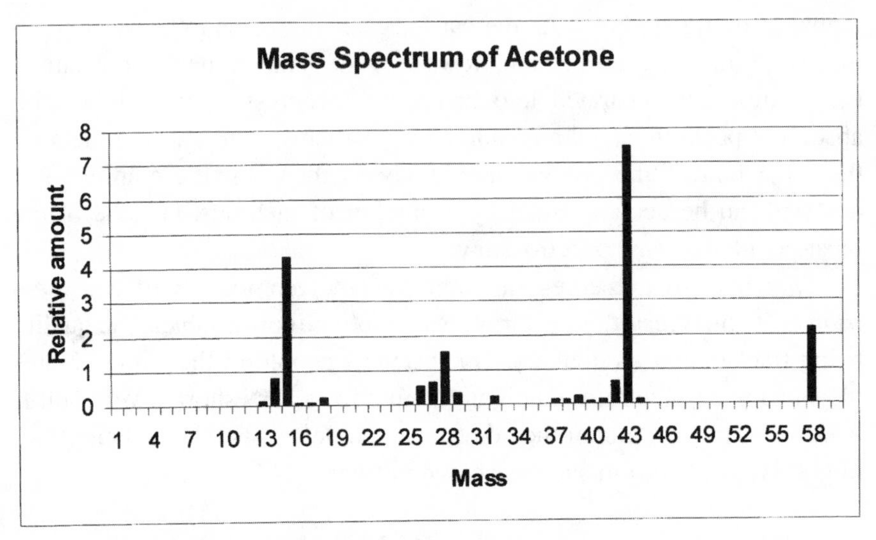


Figure 2.4.2. A mass spectrum of acetone. Molecules of acetone are broken up by a stream of electrons. The amount and relative mass of each fragment is measured. From this information and by comparing the fragment pattern with known patterns, the identity of the molecule can be established.

Mass is a pretty good identifying property. If you had one round object that weighed ten pounds (5 kilograms) and another that weighed one ounce (28 grams), you would quickly be able to sort out the shot put from the tennis ball. But the situation for mass spectroscopy is not so straightforward. In order to separate the molecules according to mass, they have to be turned into ions first, and the method for turning molecules into ions can be shattering. The molecules are hit by a stream of electrons as soon as they emerge from the chromatography column. Here again we see the power of electrons. We stated earlier that electrons are to molecules as fleas are to an elephant, but imagine splitting up our elephant with a powerful stream of fleas! Electrons may be small, but they have the power to cause chemistry.

The molecular fragments are generally missing electrons, so they have a net charge. Just as we saw with the electromagnet that we assembled from wire, a nail, and a battery, the electric field of the moving charge interacts with a magnetic field. The fragments are made to travel in a curved path by a magnetic field, with the radius of the curve being determined by the mass and the charge on the fragment. Thus the fragments are sorted by mass.

To interpret this spectrum, as with all of analytical chemistry, a little clever detective work has to be involved. The analytical chemist will look for the most massive fragment, or the *parent peak*, because this is the one that is the closest to the original mass, perhaps missing only one or two electrons or hydrogen atoms. The parent peak is not necessarily the most abundant but will have the most mass. The *cracking pattern*, or the manner in which the molecule breaks up under a stream of electrons, will depend on the relative frailty of the individual bonds. Because of this, the chemist considers the fragmenting pattern because the same type of molecule will tend to shatter in the same manner under similar conditions. Libraries of shattering patterns are maintained on computer and can be compared with the unknown. In this case, the parent peak, the one with the highest value for mass, is the peak around mass fifty-eight. Acetone has the formula $C_2H_6CO$. By consulting the appendix, we can see that an acetone molecule should have a mass of about fifty-eight: two carbon atoms at 12.01 amu plus six hydrogen atoms at 1.00 amu plus another carbon atom at 12.01 amu plus an atom of oxygen at 16.00 amu. So we have pretty good evidence that our compound is acetone. But we need corroborating evidence to be sure. A compound with the formula $C_3H_8N$ would have a mass of fifty-eight, too.

At the end of this test, the chemist hopes that the run was a good one and that the interpretation was accurate—or that there is enough sample for another try—because this technique is a *destructive technique*: the sample is now gone and cannot be reconstructed. Fortunately, there are other spectroscopic methods that are not as destructive. One of these is light spectroscopy.

As we saw when we exposed bleach to a black light, different types of light interact with matter differently, and the type of light a material interacts with can be characteristic of the material. This light-matter interaction is exploited in light spectroscopy. A sample of the material is exposed to a light beam that has been separated into components the way a prism will separate light into colors. A certain band of light is allowed to impinge on the sample and then the amount of light that makes it through the sample is detected on the other side and recorded. A new frequency of light is allowed to shine on the sample, and then the amount of this light that makes it through is again recorded, and so on. Of course, if the process were as stepwise and tedious as we have outlined, then light spectroscopy would be a very laborious technique, which, in fact, it was at one time. But many improvements later, spectroscopy is now a very fast and accurate technique.

The types of light spectroscopy commonly employed include ultraviolet spectroscopy, infrared spectroscopy, and microwave spectroscopy, though the last is less common in the forensic lab. Ultraviolet spectroscopy is often used to analyze inorganic compounds, while infrared spectroscopy is often used to analyze organic compounds. In fact, some of the terminology used for infrared spectroscopy again reminds us of our forensic chemist: the region where the absorption spectrum shows the features that are uniquely characteristic of the compound being analyzed is called the "fingerprint" region, a concept our forensic chemist should be comfortable with!

However, light spectroscopy, like mass spectroscopy, does not print out the name of the compound on a slip of paper and hand it to you. The way a molecule interacts with light is going to depend on the nature and arrangement of all of its pieces, so the response to light is varied and complicated. The analytical chemist can again resort to a library, but if the separation was not exactly clean and there are contaminants with their own, interfering, peaks, then more detective work remains to be done.

Inorganic compounds are generally less delicate than organic compounds (salts, for instance, do not burn as ethanol will) and are generally easier to analyze by nondestructive techniques. For instance, a classic analysis for an inorganic ion is to see what will make it precipitate. Carbonate ions from baking soda form a precipitate with copper ions, while the sulfate and nitrate ions in solution do not. Atomic Absorption spectroscopy (referred to around the lab as AA) is perhaps a more sensitive inorganic-analysis technique in skilled hands, but it is a destructive technique. In AA, a solution containing the analyte is *atomized*; that is, it is made into a fine mist, just as in an atomizer on a perfume bottle or a bottle of window cleaner. Once atomized, the solution to be analyzed is sprayed into an acetylene flame that has been carefully adjusted so that it has the proper temperature.

The sample needs to be introduced into the right place in the flame.

Flames themselves have a fascinating chemistry in that some regions of a flame are hotter than others and some regions promote oxidation reactions while others promote reduction. And there is, of course, the most interesting aspect of flame chemistry: flames produce heat and light. The color of the light will depend on the material that is in the flame, just as the color of fireworks depends on the material in the ordnance. In one configuration, the AA instrument takes advantage of this fact to analyze the material that has been atomized into the flame; a light detector near the flame records the different colors of the flame and their intensity, which is proportional to the amount of each material in the flame.

In another configuration, an AA instrument uses other commonly exploited physical phenomena: the fact that materials tend to absorb the same frequency of light as they emit. Rose-colored glasses make the world rosier because the materials from which the lens are made absorb all the colors except rose, and materials can be identified by the colors they absorb. In AA, light from a lamp designed to emit a very narrow range of color is directed onto the sample in the flame. The lamp is designed so that the material for analysis is also the material producing the light in the lamp. In other words, if the analyst wants to analyze for sodium, then a sodium-containing lamp is used. The lamp then emits a light that is characteristic of sodium, and if there is any sodium in the sample, it will absorb this light and the detector will see that some of the light is missing.

The methods employed by our intrepid forensic chemist are infrared and mass spectroscopy. At the end of her efforts, she has found that indeed there are traces of peanut oil on the victim's mouth, so the questions become this: Where did it come from? Where did it go? Our heroine analyzes the samples of foods and medicines she took from the apartment but finds nothing. She notes that the victim must have read labels very carefully so as to avoid even seemingly innocent products such as some ice creams, baked goods, salad dressings, and soups that could contain peanut oil. The stomach and blood showed no other toxins, though there is evidence of a severe allergic reaction. She is pondering this thought and carefully cataloging her results when the phone rings. The officers are calling to tell her she was right. They found what she told them to look for: an unused epipen—a syringe full of lifesaving medicine—discarded in a dumpster behind the apartment building along with a pair of surgical gloves.

How did our chemist know to tell the officers to seach for an epipen outside the apartment? When she looked in the cupboard, the medicine cabinet, and on the victim she didn't find an epipen. She knew that the victim had been very careful about picking her foods to be allergen free, so the victim was aware of her allergy. Anyone who was aware she had such a dangerous allergy would most certainly have an epipen where she could get to it—unless she were prevented from getting to it. Our forensic chemist now knows that on the surface of the gloves she will find peanut oil . . . and a nice set of fingerprints inside!

So the job of the analytical chemist may be complex, difficult, demanding, and exacting—but it still pays better than crime. And though certain people may have sensitivities to some plants, and though some plants may be poisonous to all people, the interactions between plants and animals are not always so adversarial. Plants can sometimes be very beneficial. In fact, when it comes to plants and animals, they have a future together—the future of chemistry.

## Demonstration for the Future: Medicine Cabinet Chemistry

For our last demonstration, go to your medicine cabinet and take a synthetic mimic of the extract of Salix willow bark, modified to buffer the acidic effect. In other words, take an aspirin.

What does aspirin have to do with the future of chemistry? Read on.

# CHAPTER 5

## Harry, Hogwarts, and Folk Pharmacopoeias—Mysteries in the Past and Magic in the Future

*"You are here to learn the subtle science and exact art of potion-making. . . . As there is little foolish wand-waving here, many of you will hardly believe this is magic. . . . (T)he beauty of the softly simmering cauldron with its shimmering fumes, the delicate power of liquids that creep through human veins, bewitching the minds, ensnaring the senses . . . I can teach you how to bottle fame, brew glory, even stopper death— . . ."*

*"Potter!" said Snape suddenly. "What would I get if I added powdered root of asphodel to an infusion of wormwood?"*

—J. K. Rowling, *Harry Potter and the Sorcerer's Stone*, ca. 1997

*Do you believe then that the sciences would ever have arisen and become great if there had not beforehand been magicians, alchemists, astrologers, and wizards who thirsted and hungered after . . . forbidden powers?*

—Friedrich Wilhelm Nietzsche, ca. 1890

There are tens of millions of known chemical compounds and more being assembled every day. The number of future combinations of elements is virtually infinite, and the periodic table continues to expand. We've become almost too good at our craft. Like kids in a candy store, the question is not whether there is more chemistry to be discovered but where we should focus first. With all the problems facing human survival, we can't afford to sit like a million monkeys working in a million labs and hope that solutions will arise. Take, for example, the problem of drug discovery. How do we decide, out of the tens of millions of compounds out there, which medicines to test as cures for diseases? In the past, there have been some fortunate accidents, but we can't rely on luck.

In the process of rational drug design, there are a couple of choices. First, we can try variations on compounds that have worked well in the past. This approach is reasonable, respectable, and often successful. When a particular type of bacterium evolves a strain that is immune to our medicines, we can take a medicine that worked previously and tack on a new group of atoms or change a bond to see if we can make it effective again. This approach, called the *combinatorial approach*, has yielded many useful compounds. But the approach has also been likened to a tree with many roots: it can sample a lot of soil, but only soil in the vicinity of the tree. If there is a new drug with a completely new structure, then the combinatorial approach may not find it.

Another approach that has been successful in the past, whether or not by design, is to let the millions of years of human-plant interaction do the work of the millions of tests. Aspirin was "discovered" when someone decided to see if there was anything to the old folk remedy of chewing on willow bark to relieve pain and of drinking willow bark tea to reduce fever.[1] Quinine was "discovered" when a desperate monk gave a dying noble woman an extract of bark recommended to him by the indigenous people of Peru.[2] There is currently serious and intense research being directed toward the study of folk medicines, especially medicines derived from plants.

J. K. Rowling's popular Harry Potter series no doubt owes its success to her enchanting characters, intriguing stories, clever dialogue, and engaging setting.[3] But there may be another factor, too. The references to magic and magical herbs, stones, and fungi resonate with a cultural memory: we find it easy to believe in magic because our forbearers experienced what appeared to be magic. Many of the "magical" herbs mentioned in the Harry Potter series, such as wormwood, named for its power for expunging intestinal worms, have real efficacy.[4] Imagine suffering with intestinal worms and finding a preparation of wormwood could rid you of the parasites. Imagine being plagued with goiter and then finding relief in iodine-containing charred seaweed. Imagine suffering from scurvy and finding respite in the juice of limes. These were all real cures that were used in the past and, to the sufferer, had to seem magical.

Take another look in your medicine cabinet to see other products that had their origin in plant remedies: beta blockers, birth control, sedatives, laxatives, antiseptics, soap. . . . So the next time you take a much-needed medicine, go out and hug a tree.

The study of the chemistry of plants is termed *phytochemistry*, and the fact that a number of effective drugs have been found in natural plant products (some say as many as a third of the current manufactured drugs) begs the question: Why? Why should it be that plants produce materials that interact with humans? Why should certain plant products make us healthy, some make us happy, some make us sleepy, and some make us dead?

Certainly it could be coincidence. We evolved on the same planet—humans and plants—so our chemistries should be similar. Medicines of plant origin are more likely to be compatible with our systems because we share the same chemical space. But there is also a well-known plant-animal mutualism: what's good for the gooseberry is good for the goose. Plants consume carbon dioxide and produce oxygen; animals consume oxygen and produce carbon dioxide. Bees pollinate flowers; flowers feed bees. The most ferocious, carnivorous cat will occasionally munch grass for the fiber it needs. Ovaries of some plants grow into fruit so animals will eat the seed and carry it away—thus avoiding competition with the parent plant for resources.

Some chemicals that plants produce are useful to us because plants protect themselves from the same things we do—bacteria, viruses, and fungi. But herbs can do more than act as antiseptics. As Snape scolds Potter, Harry needs to know his herbs: asphodel, an herb, is used to "make a sleeping potion so powerful it is known as the Draught of Living Death."[5] And it is true.[6] Harry is cautioned that monkshood and wolfsbane are effective poisons.[7] And they are.[8] Snape instructs Harry to copy down the name "aconite"—and history tells us that aconite and belladonna were once used together to produce the hallucination of flying.[9]

Why would plants produce chemicals with these peculiar properties? Do plants have headaches? Do they have trouble sleeping? Why have they evolved chemicals with analgesic, sedative, or psychoactive properties?

Again, it could be just an accident. Plants evolved chemicals to kill predators, but a toxin that would kill an insect might just tantalize or sedate a human. The difference between a sedative and a poison is often a matter of dosage. The ingestion of toxins below lethal levels for the sake of euphoric effects has become known as *intoxication*, as a nod to its toxic origins.

On the other hand, plants may have developed some sedatives and analgesics because there are advantages to the plant. If a plant is being eaten for food, there may be no incentive for the predator to restrain itself from consuming the entire growth, but if the plant were being eaten for the sensory benefit, the animal might eat only enough to bring on the effect and then stop, stupefied. What would be the benefit to the plant? Maybe fertilization. An animal that is sedated by a plant may be more likely to defecate in the plant's vicinity instead of wandering away with the plant's sacrificed nutrients. Or the advantage for the plant might be pruning. Herbaceous plants can propagate by leaf cuttings and seed. Pinching off the tops of these types of plants allows the plant to concentrate it efforts on the remaining growth. If there were a sedative in the plant, then the pruning would be self-limiting. The same chemical that acts as a sedative for animals might be a poison for bugs, which would limit the insect pruning, too.

From whatever evolutionary pressures, however, plants have evolved compounds that have an effect on us. Coffee, tea, tobacco, licorice, codeine, cocaine, and opium—all of these were originally herbal cures. Plants are sources of effective antiseptics, analgesics, antibiotics, vermifuges, sedatives, stimulants, abortifacients, poisons, laxatives, diuretics, antidiarrheals, and, yes, recreational drugs. As Harry Potter's potions master says, bottled fame, brewed glory, and even stoppered death.

*Pharmacognosy* is the study of natural product drug discovery and development, and pharmacognosy includes in its scope the investigation of ethno-botanical claims—in other words, folk medicine. These researchers look at practices of modern cultures, such as the inhabitants of the rainforest, and they also look at historical practitioners, such as the monks and mendicants of the European Middle Ages—and the old wives, the wizards, and the witches.

There is a new realization that many of the people who were persecuted as witches and sorcerers in the European Middle Ages were, in fact, innocent midwives and healers, and some of their so-called magical remedies had certifiable efficacy. They recommended analgesic clove oil for toothache, vitamin-containing herbal teas for pregnant women and listless children, and calcined egg shell for upset stomach, a source of calcium carbonate, a main ingredient in antacids prescribed today.

But not all of the medical practices of the ancients were useful. They did not follow standard, sound scientific procedures such as keeping records of successful trials and failure, and they did not carefully sort out cause and effect. Their medical philosophies were known to include treating the weapon that caused the wound instead of the wound and using medicines with effects that resembled the disease that they were meant to cure—which could be the origin of taking a "hair of the dog that bit you" as a cure. The Arabs brought a lot of practical medical knowledge, as well as the knowledge of the ancient Greeks and Romans, to the Western Europeans in the early Middle Ages. Consequently, European healers started imitating Arabic speech in hopes of improving their own efforts, hence "magic" words like alakazam and abracadabra. J. K. Rowling captures the notion well when she has her apprentice sorcerers chant their spells in erudite Latin-sounding magical phases.[10]

But sometimes the ancient healers *did* know what they were doing—but didn't want anyone else to know. The European Middle Ages predated patent medicine by many years, so successful healers had to guard their secrets if they wanted to keep their jobs. Many of the medicines that they used were fairly recognizable, such as aromatic herbs, so they would disguise the actual active ingredient by serving it in a foul-tasting and foul-smelling concoction. They constructed smoke screens of elaborate rituals so that the administration of the medicine was only one of many more impressive steps. It is no accident the novice Harry Potter "had been really looking forward to" his first magic wand.[11] Other practitioners of magic were known to wave a stick around so that the patient would be watching the stick and not see what they were slipping into the stew. The hocus-pocus might also have acted as a sort of primitive malpractice insurance. If the cure was ineffective, the witch or wizard could always claim that their instructions hadn't been followed. Or that a certain herb should have been picked at midnight during a full moon.

If there is so much nonsense, how do we separate the useful herbs from the hooey? Or can we? Is there any real point in examining ethnic medical traditions? Yes. Studying how the herbs are used now—and how they were used in the past—tells us more than just grinding up plants and running them through a mass spectrometer. It is important to know how the herb is prepared when evaluating its efficacy. The proper preparation may be essential to release the active ingredient. It is also necessary to know how the herb was presented. Many times drugs are synergistic, only working in combination with other drugs, an effect that could not be discerned by testing them separately. The possibility of interactions is also the reason it is imperative to check with a doctor when combining any drugs, even herbal preparations. Sometimes very adverse interactions can occur. The wonderful relief from intestinal parasites provided by wormwood could easily turn into disaster or even death: wormwood, along with other herbs, is used to prepare absinthe, a notoriously addictive liqueur known to cause blindness, nerve damage, and mental instability in habitual users.[12] Investigations into the traditional manner in which an herb was used can provide information that might otherwise be overlooked.

As well as investigating folk pharmacopeias, pharmacognosy looks for new drugs from natural sources, too, and there is the promise of many interesting chemicals to be found. There are trees that live thousands of years. What do they know that we don't? And when it comes to growing limbs, trees have it all over us. Under attack by insect pests, plants have been found to send out chemical signals to attract other insects that feed on the harmful bugs. Recently, it was shown that some plants under attack can even send out warning signals to other plants to arm themselves against impending danger.[13] There may be a wealth of other such chemical compounds in plants that are only present part of the time, such as when the plant is under duress or in certain atmospheric conditions. Such as midnight under a full moon?

There are no doubt other useful materials elsewhere in nature, too: in the oceans, polar icecaps, volcanoes, and maybe even other planets and moons. Before we can take advantage of these resources, there are a host of problems to be solved—but the possibility of a chemical cornucopia if we solve them. Where do we focus our efforts? The answer may be found in the words of the venerable scientist and science writer, Isaac Asimov. He said, "The most exciting phrase to hear in science . . . is not 'Eureka!' but 'That's funny . . .'"

We will find our direction when a discovery is made. When someone picks up a pebble and notes a new pattern. Or finds a new insect or an odd new ore. It will take someone with curiosity and a desire to know. Someone who enjoys history, music, and art—and perhaps peculiar chemistry books.

# EPILOGUE

*I celebrate myself, and sing myself,*
*And what I assume you shall assume,*
*For every atom belonging to me as good belongs to you.*

—Walt Whitman, *Leaves of Grass,* 1855

If all our readers, provoked by this too-brief outline of the science of chemistry, should arise as one and storm out in quest of further edification—nothing could please us more. There is so much more to be told.

But we've been on an excellent journey. We have examined the structure and behavior of minute particles of matter, and we have seen how their microscopic properties multiply to form the world around us. We surveyed classical chemistry and demonstrated these lively reactions with ordinary materials. We have seen how chemistry relates to everyday experience, and we've developed an intuitive understanding of the essentials of chemistry by example and analogy. We've seen how principles and theory derived in the past have showered us with a wealth of technology and materials—and we have seen the challenges that lie ahead. In this new millennium, there will be new medicines and innovative approaches to healing. There will be bridges that will bend, not break, when the earth quakes. There will be new sources of fuels as abundant as the sunlight and enough food to feed a hungry world. There will be a new understanding of all the sciences, and we will learn how to protect this planet and life in all its forms.

But where do we start? Who determines where to apply our resources? If we find a new life form in the ocean that produces a potent new drug, to whom does it belong? Who would be allowed to harvest it? Develop it? Ensure its preservation? People. All people. The chemistry of the future is going to depend on more than chemists. It's going to require international cooperative efforts that draw on the talents of philosophers and prospectors; ecologists and ethicists; economists and statisticians; lawyers and law enforcement; botanists and bacteriologists; herbalists and historians; authors and their audiences; laboratory, electrical, and computer technologists; and people ready to make educated, informed decisions.

There is so much work to be done, and so many reasons to do it.

The work is far from finished. And the joy never ends.

# APPENDIX

## Elements Listed by Name, Symbol, and Atomic Number

### Elements Sorted by Name

| Symbol | Name | Atomic Number | Mass* |
|---|---|---|---|
| Ac | Actinium | 89 | (227) |
| Al | Aluminum | 13 | 26.98 |
| Am | Americium | 95 | (243) |
| Sb | Antimony | 51 | 121.76 |
| Ar | Argon | 18 | 39.95 |
| As | Arsenic | 33 | 74.92 |
| At | Astatine | 85 | (210) |
| Ba | Barium | 56 | 137.33 |
| Bk | Berkelium | 97 | (247) |
| Be | Beryllium | 4 | 9.01 |
| Bi | Bismuth | 83 | 208.98 |
| Bh | Bohrium | 107 | (264) |
| B | Boron | 5 | 10.81 |
| Br | Bromine | 35 | 79.90 |
| Cd | Cadmium | 48 | 112.41 |
| Ca | Calcium | 20 | 40.08 |
| Cf | Californium | 98 | (251) |
| C | Carbon | 6 | 12.01 |

| | | | |
|---|---|---|---|
| Ce | Cerium | 58 | 140.12 |
| Cs | Cesium | 55 | 132.91 |
| Cl | Chlorine | 17 | 35.45 |
| Cr | Chromium | 24 | 52.00 |
| Co | Cobalt | 27 | 58.93 |
| Cu | Copper | 29 | 63.55 |
| Cm | Curium | 96 | (247) |
| Db | Dubnium | 105 | (262) |
| Dy | Dysprosium | 66 | 162.50 |
| Es | Einsteinium | 99 | (252) |
| Er | Erbium | 68 | 167.26 |
| Eu | Europium | 63 | 151.96 |
| Fm | Fermium | 100 | (257) |
| F | Fluorine | 9 | 19.00 |
| Fr | Francium | 87 | (223) |
| Gd | Gadolinium | 64 | 157.25 |
| Ga | Gallium | 31 | 69.72 |
| Ge | Germanium | 32 | 72.64 |
| Au | Gold | 79 | 196.97 |
| Hf | Hafnium | 72 | 178.49 |
| Hs | Hassnium | 108 | (269) |
| He | Helium | 2 | 4.00 |
| Ho | Holmium | 67 | 164.93 |
| H | Hydrogen | 1 | 1.01 |
| In | Indium | 49 | 114.82 |
| I | Iodine | 53 | 126.90 |
| Ir | Iridium | 77 | 192.22 |
| Fe | Iron | 26 | 55.85 |
| Kr | Krypton | 36 | 83.80 |
| La | Lanthanum | 57 | 138.91 |
| Lr | Lawrencium | 103 | (262) |
| Pb | Lead | 82 | 207.20 |
| Li | Lithium | 3 | 6.94 |
| Lu | Lutetium | 71 | 174.97 |
| Mg | Magnesium | 12 | 24.31 |
| Mn | Manganese | 25 | 54.94 |
| Mt | Meitnerium | 109 | (268) |
| Md | Mendelevium | 101 | (258) |
| Hg | Mercury | 80 | 200.59 |

| | | | |
|---|---|---|---|
| Mo | Molybdenum | 42 | 95.94 |
| Nd | Neodymium | 60 | 144.24 |
| Ne | Neon | 10 | 20.12 |
| Np | Neptunium | 93 | 237.05 |
| Ni | Nickel | 28 | 58.69 |
| Nb | Niobium | 41 | 92.91 |
| N | Nitrogen | 7 | 14.01 |
| No | Nobelium | 102 | (259) |
| Os | Osmium | 76 | 190.23 |
| O | Oxygen | 8 | 16.00 |
| Pd | Palladium | 46 | 106.42 |
| P | Phosphorus | 15 | 30.97 |
| Pt | Platinum | 78 | 195.08 |
| Pu | Plutonium | 94 | (244) |
| Po | Polonium | 84 | (209) |
| K | Potassium | 19 | 39.10 |
| Pr | Praseodymium | 59 | 140.91 |
| Pm | Promethium | 61 | (145) |
| Pa | Protactinium | 91 | 231.04 |
| Ra | Radium | 88 | (226) |
| Rn | Radon | 86 | (222) |
| Re | Rhenium | 75 | 186.21 |
| Rh | Rhodium | 45 | 102.91 |
| Rb | Rubidium | 37 | 85.47 |
| Ru | Ruthenium | 44 | 101.07 |
| Rf | Rutherfordium | 104 | (261) |
| Sm | Samarium | 62 | 150.36 |
| Sc | Scandium | 21 | 44.96 |
| Sg | Seaborgium | 106 | (266) |
| Se | Selenium | 34 | 78.96 |
| Si | Silicon | 14 | 28.09 |
| Ag | Silver | 47 | 107.87 |
| Na | Sodium | 11 | 22.99 |
| Sr | Strontium | 38 | 87.62 |
| S | Sulfur | 16 | 32.07 |
| Ta | Tantalum | 73 | 180.95 |
| Tc | Technetium | 43 | (98) |
| Te | Tellurium | 52 | 127.60 |
| Tb | Terbium | 65 | 158.93 |

| | | | |
|---|---|---|---|
| Tl | Thallium | 81 | 204.38 |
| Th | Thorium | 90 | 232.04 |
| Tm | Thulium | 69 | 168.93 |
| Sn | Tin | 50 | 118.71 |
| Ti | Titanium | 22 | 47.87 |
| W | Tungsten | 74 | 183.84 |
| U | Uranium | 92 | 238.03 |
| V | Vanadium | 23 | 50.94 |
| Xe | Xenon | 54 | 131.29 |
| Yb | Ytterbium | 70 | 173.04 |
| Y | Yttrium | 39 | 88.91 |
| Zn | Zinc | 30 | 65.39 |
| Zr | Zirconium | 40 | 91.22 |

*The masses can be interpreted as amu per atom (see chap. 2) or grams per mole (see chap. 9). The numbers in parentheses are approximate because these elements are artificially made. They do not have a natural abundance on which to base an average mass.

## ELEMENTS SORTED BY SYMBOL

| Symbol | Name | Atomic Number | Mass* |
|---|---|---|---|
| Ac | Actinium | 89 | (227) |
| Ag | Silver | 47 | 107.87 |
| Al | Aluminum | 13 | 26.98 |
| Am | Americium | 95 | (243) |
| Ar | Argon | 18 | 39.95 |
| As | Arsenic | 33 | 74.92 |
| At | Astatine | 85 | (210) |
| Au | Gold | 79 | 196.97 |
| B | Boron | 5 | 10.81 |
| Ba | Barium | 56 | 137.33 |
| Be | Beryllium | 4 | 9.01 |
| Bh | Bohrium | 107 | (264) |
| Bi | Bismuth | 83 | 208.98 |
| Bk | Berkelium | 97 | (247) |
| Br | Bromine | 35 | 79.90 |
| C | Carbon | 6 | 12.01 |
| Ca | Calcium | 20 | 40.08 |
| Cd | Cadmium | 48 | 112.41 |
| Ce | Cerium | 58 | 140.12 |
| Cf | Californium | 98 | (251) |
| Cl | Chlorine | 17 | 35.45 |
| Cm | Curium | 96 | (247) |
| Co | Cobalt | 27 | 58.93 |
| Cr | Chromium | 24 | 52.00 |
| Cs | Cesium | 55 | 132.91 |
| Cu | Copper | 29 | 63.55 |
| Db | Dubnium | 105 | (262) |
| Dy | Dysprosium | 66 | 162.50 |
| Er | Erbium | 68 | 167.26 |
| Es | Einsteinium | 99 | (252) |
| Eu | Europium | 63 | 151.96 |
| F | Fluorine | 9 | 19.00 |
| Fe | Iron | 26 | 55.85 |
| Fm | Fermium | 100 | (257) |
| Fr | Francium | 87 | (223) |
| Ga | Gallium | 31 | 69.72 |

| | | | |
|---|---|---|---|
| Gd | Gadolinium | 64 | 157.25 |
| Ge | Germanium | 32 | 72.64 |
| H | Hydrogen | 1 | 1.01 |
| He | Helium | 2 | 4.00 |
| Hf | Hafnium | 72 | 178.49 |
| Hg | Mercury | 80 | 200.59 |
| Ho | Holmium | 67 | 164.93 |
| Hs | Hassnium | 108 | (269) |
| I | Iodine | 53 | 126.90 |
| In | Indium | 49 | 114.82 |
| Ir | Iridium | 77 | 192.22 |
| K | Potassium | 19 | 39.10 |
| Kr | Krypton | 36 | 83.80 |
| La | Lanthanum | 57 | 138.91 |
| Li | Lithium | 3 | 6.94 |
| Lr | Lawrencium | 103 | (262) |
| Lu | Lutetium | 71 | 174.97 |
| Md | Mendelevium | 101 | (258) |
| Mg | Magnesium | 12 | 24.31 |
| Mn | Manganese | 25 | 54.94 |
| Mo | Molybdenum | 42 | 95.94 |
| Mt | Meitnerium | 109 | (268) |
| N | Nitrogen | 7 | 14.01 |
| Na | Sodium | 11 | 22.99 |
| Nb | Niobium | 41 | 92.91 |
| Nd | Neodymium | 60 | 144.24 |
| Ne | Neon | 10 | 20.12 |
| Ni | Nickel | 28 | 58.69 |
| No | Nobelium | 102 | (259) |
| Np | Neptunium | 93 | 237.05 |
| O | Oxygen | 8 | 16.00 |
| Os | Osmium | 76 | 190.23 |
| P | Phosphorus | 15 | 30.97 |
| Pa | Protactinium | 91 | 231.04 |
| Pb | Lead | 82 | 207.20 |
| Pd | Palladium | 46 | 106.42 |
| Pm | Promethium | 61 | (145) |
| Po | Polonium | 84 | (209) |
| Pr | Praseodymium | 59 | 140.91 |

| | | | |
|---|---|---|---|
| Pt | Platinum | 78 | 195.08 |
| Pu | Plutonium | 94 | (244) |
| Ra | Radium | 88 | (226) |
| Rb | Rubidium | 37 | 85.47 |
| Re | Rhenium | 75 | 186.21 |
| Rf | Rutherfordium | 104 | (261) |
| Rh | Rhodium | 45 | 102.91 |
| Rn | Radon | 86 | (222) |
| Ru | Ruthenium | 44 | 101.07 |
| S | Sulfur | 16 | 32.07 |
| Sb | Antimony | 51 | 121.76 |
| Sc | Scandium | 21 | 44.96 |
| Se | Selenium | 34 | 78.96 |
| Sg | Seaborgium | 106 | (266) |
| Si | Silicon | 14 | 28.09 |
| Sm | Samarium | 62 | 150.36 |
| Sn | Tin | 50 | 118.71 |
| Sr | Strontium | 38 | 87.62 |
| Ta | Tantalum | 73 | 180.95 |
| Tb | Terbium | 65 | 158.93 |
| Tc | Technetium | 43 | (98) |
| Te | Tellurium | 52 | 127.60 |
| Th | Thorium | 90 | 232.04 |
| Ti | Titanium | 22 | 47.87 |
| Tl | Thallium | 81 | 204.38 |
| Tm | Thulium | 69 | 168.93 |
| U | Uranium | 92 | 238.03 |
| V | Vanadium | 23 | 50.94 |
| W | Tungsten | 74 | 183.84 |
| Xe | Xenon | 54 | 131.29 |
| Y | Yttrium | 39 | 88.91 |
| Yb | Ytterbium | 70 | 173.04 |
| Zn | Zinc | 30 | 65.39 |
| Zr | Zirconium | 40 | 91.22 |

*The masses can be interpreted as amu per atom (see chap. 2) or grams per mole (see chap. 9). The numbers in parentheses are approximate because these elements are artificially made. They do not have a natural abundance on which to base an average mass.

## Elements Sorted by Atomic Number

(the periodic table is also a list of elements by atomic number)

| Symbol | Name | Atomic Number | Mass* |
|---|---|---|---|
| H | Hydrogen | 1 | 1.01 |
| He | Helium | 2 | 4.00 |
| Li | Lithium | 3 | 6.94 |
| Be | Beryllium | 4 | 9.01 |
| B | Boron | 5 | 10.81 |
| C | Carbon | 6 | 12.01 |
| N | Nitrogen | 7 | 14.01 |
| O | Oxygen | 8 | 16.00 |
| F | Fluorine | 9 | 19.00 |
| Ne | Neon | 10 | 20.12 |
| Na | Sodium | 11 | 22.99 |
| Mg | Magnesium | 12 | 24.31 |
| Al | Aluminum | 13 | 26.98 |
| Si | Silicon | 14 | 28.09 |
| P | Phosphorus | 15 | 30.97 |
| S | Sulfur | 16 | 32.07 |
| Cl | Chlorine | 17 | 35.45 |
| Ar | Argon | 18 | 39.95 |
| K | Potassium | 19 | 39.10 |
| Ca | Calcium | 20 | 40.08 |
| Sc | Scandium | 21 | 44.96 |
| Ti | Titanium | 22 | 47.87 |
| V | Vanadium | 23 | 50.94 |
| Cr | Chromium | 24 | 52.00 |
| Mn | Manganese | 25 | 54.94 |
| Fe | Iron | 26 | 55.85 |
| Co | Cobalt | 27 | 58.93 |
| Ni | Nickel | 28 | 58.69 |
| Cu | Copper | 29 | 63.55 |
| Zn | Zinc | 30 | 65.39 |
| Ga | Gallium | 31 | 69.72 |
| Ge | Germanium | 32 | 72.64 |
| As | Arsenic | 33 | 74.92 |
| Se | Selenium | 34 | 78.96 |
| Br | Bromine | 35 | 79.90 |

| | | | |
|---|---|---|---|
| Kr | Krypton | 36 | 83.80 |
| Rb | Rubidium | 37 | 85.47 |
| Sr | Strontium | 38 | 87.62 |
| Y | Yttrium | 39 | 88.91 |
| Zr | Zirconium | 40 | 91.22 |
| Nb | Niobium | 41 | 92.91 |
| Mo | Molybdenum | 42 | 95.94 |
| Tc | Technetium | 43 | (98) |
| Ru | Ruthenium | 44 | 101.07 |
| Rh | Rhodium | 45 | 102.91 |
| Pd | Palladium | 46 | 106.42 |
| Ag | Silver | 47 | 107.87 |
| Cd | Cadmium | 48 | 112.41 |
| In | Indium | 49 | 114.82 |
| Sn | Tin | 50 | 118.71 |
| Sb | Antimony | 51 | 121.76 |
| Te | Tellurium | 52 | 127.60 |
| I | Iodine | 53 | 126.90 |
| Xe | Xenon | 54 | 131.29 |
| Cs | Cesium | 55 | 132.91 |
| Ba | Barium | 56 | 137.33 |
| La | Lanthanum | 57 | 138.91 |
| Ce | Cerium | 58 | 140.12 |
| Pr | Praseodymium | 59 | 140.91 |
| Nd | Neodymium | 60 | 144.24 |
| Pm | Promethium | 61 | (145) |
| Sm | Samarium | 62 | 150.36 |
| Eu | Europium | 63 | 151.96 |
| Gd | Gadolinium | 64 | 157.25 |
| Tb | Terbium | 65 | 158.93 |
| Dy | Dysprosium | 66 | 162.50 |
| Ho | Holmium | 67 | 164.93 |
| Er | Erbium | 68 | 167.26 |
| Tm | Thulium | 69 | 168.93 |
| Yb | Ytterbium | 70 | 173.04 |
| Lu | Lutetium | 71 | 174.97 |
| Hf | Hafnium | 72 | 178.49 |
| Ta | Tantalum | 73 | 180.95 |
| W | Tungsten | 74 | 183.84 |

| | | | |
|---|---|---|---|
| Re | Rhenium | 75 | 186.21 |
| Os | Osmium | 76 | 190.23 |
| Ir | Iridium | 77 | 192.22 |
| Pt | Platinum | 78 | 195.08 |
| Au | Gold | 79 | 196.97 |
| Hg | Mercury | 80 | 200.59 |
| Tl | Thallium | 81 | 204.38 |
| Pb | Lead | 82 | 207.20 |
| Bi | Bismuth | 83 | 208.98 |
| Po | Polonium | 84 | (209) |
| At | Astatine | 85 | (210) |
| Rn | Radon | 86 | (222) |
| Fr | Francium | 87 | (223) |
| Ra | Radium | 88 | (226) |
| Ac | Actinium | 89 | (227) |
| Th | Thorium | 90 | 232.04 |
| Pa | Protactinium | 91 | 231.04 |
| U | Uranium | 92 | 238.03 |
| Np | Neptunium | 93 | 237.05 |
| Pu | Plutonium | 94 | (244) |
| Am | Americium | 95 | (243) |
| Cm | Curium | 96 | (247) |
| Bk | Berkelium | 97 | (247) |
| Cf | Californium | 98 | (251) |
| Es | Einsteinium | 99 | (252) |
| Fm | Fermium | 100 | (257) |
| Md | Mendelevium | 101 | (258) |
| No | Nobelium | 102 | (259) |
| Lr | Lawrencium | 103 | (262) |
| Rf | Rutherfordium | 104 | (261) |
| Db | Dubnium | 105 | (262) |
| Sg | Seaborgium | 106 | (266) |
| Bh | Bohrium | 107 | (264) |
| Hs | Hassnium | 108 | (269) |
| Mt | Meitnerium | 109 | (268) |

*The masses can be interpreted as amu per atom (see chap. 2) or grams per mole (see chap. 9). The numbers in parentheses are approximate because these elements are artificially made. They do not have a natural abundance on which to base an average mass.

# ENDNOTES AND CREDITS

## CREDITS

Original artwork by Linda Muse appears on pages 15, 17, 27, 29, 33, 69, 73, 83, 95, 114, 115, 117, 119, 123, 131, 141, 182, 187, 198, 212, 233, 241, 247, 271, 273, 282, 287, 311, and 315.

## ENDNOTES

### Apologia

1. Kerry K. Karukstis and Gerald R. Van Hecke, *Chemistry Connections: The Chemical Basis of Everyday Phenomena* (New York: Harcourt/Academic Press, 2000); Joe Schwarcz, *The Genie in the Bottle* (New York: W. H. Freeman, 2001).

### A Few Necessary Words on Safety

1. For example, Robert Gardner, *Kitchen Chemistry: Science Experiments To Do at Home* (New York: J. Messner, 1982); Alan Kramer, *How To Make a Chemical Volcano and Other Mysterious Experiments* (New York: F. Watts, 1989); Louis V. Loeschnig, *Simple Chemistry Experiments with Everyday Materials* (New York: Sterling, 1994); Nathan Shalit, *Cup and Saucer Chemistry* (New York: Dover, 1989).

## Introduction

1. Dr. Seuss, *Bartholomew and the Oobleck* (New York: Random House, 1976).

2. "Nobel Lectures, Chemistry 1942–1962," http://www.nobel.se/chemistry/index.html.

3. Tore Frängsmyr, ed., *Les Prix Nobel 1995* (Stockholm: Almqvist & Wiksell International, 1995). Also online at http://www.nobel.se/chemistry/index.html.

4. Georgina Ferry, *Dorothy Hodgkin: A Life* (Cold Spring Harbor, NY: Cold Spring Harbor Laboratory Press, 1998), pp. 7–25.

5. Frängsmyr, *Les Prix Nobel 1996.*

## Electrons and Atoms, Elephants and Fleas

1. Paul A. Tipler, *Physics for Scientists and Engineers* (New York: Worth, 1992), pp. 598–600.

2. Lennard Bickel, *The Deadly Element: The Story of Uranium* (New York: Stein and Day, 1979), p. 66.

3. C. C. Gillispie, ed., *Dictionary of Scientific Biography*, 18 vols. (New York: Scribner, 1970–1990), 13: 362–72.

4. Mary Jo Nye, *Molecular Reality: A Perspective on the Scientific Work of Jean Perrin* (New York: American Elsevier, 1972), pp. 51–172.

5. Ruth Lewin Sime, *Journal of Chemical Education* 66 (1989): 373.

6. Linda Merricks, *The World Made New: Frederick Soddy, Science, Politics, and Environment* (New York: Oxford University Press, 1996), p. 70.

7. Gillispie, *Dictionary of Scientific Biography*, 17: 143–48.

8. Many thanks to Jack Goldsmith for making us aware of this analogy.

## Periodically Speaking

1. Mary Elvira Weeks and Henry M. Leichester, *Discovery of the Elements*, 7th ed. (Austin, TX: Journal of Chemical Education, 1968), p. 662.

2. Ira N. Levine, *Quantum Chemistry*, 3rd ed. (Boston: Allyn and Bacon, 1983), p. 242.

3. Elaine N. Marieb, *Human Anatomy and Physiology*, 3rd ed. (New York: Benjamin/Cummings, 1991), p. 28.

4. "Alzheimer's Disease," National Institute of Environmental Health Science (NIEHS), http://www.niehs.nih.gov/external/faq/aluminum.htm.

## Reason, Reactions, and Redox

1. Norman Mclean, *Young Men and Fire* (Chicago: University of Chicago Press, 1992), pp. 57–130.

## The Tie That Binds, the Chemicals That Bond

1. Lee R. Summerlin, Christie L. Borgford, and Julie B. Ealy, *Chemical Demonstrations: A Sourcebook for Teachers*, 2 vols. (Washington, DC: American Chemical Society, 1987), 2: 179.

## It's a Gas

1. J. R. Partington, *A Short History of Chemistry*, 3rd ed. (New York: Macmillan, 1957), p. 46.
2. Ibid., pp. 72–73.
3. C. C. Gillispie, ed., *Dictionary of Scientific Biography*, 18 vols. (New York: Scribner, 1970–1990), 3: 207–209.
4. Ibid., 5: 317–26.
5. Ibid., 13: 374–88.
6. Keith J. Laidler, *The World of Physical Chemistry* (New York: Oxford University Press, 1995), p. 348.

## When Gases Put On Airs

1. Donald L. Pavia, Gary M. Lampman, and George S. Kriz Jr., *Introduction to Organic Laboratory Techniques* (Philadelphia: Saunders, 1976), pp. 125–29.
2. C. L. Cobb and H. Goldwhite, *Creations of Fire: Chemistry's Lively History from Alchemy to the Atomic Age* (New York: Plenum, 1995), pp. 302–306.
3. Richard A. Kerr, "Ancient Air Analyzed in Dinosaur-Age Amber," *Science* 238 (November 13, 1987): 890.

## Crystal Clear Chemistry

1. C. L. Cobb and H. Goldwhite, *Creations of Fire: Chemistry's Lively History from Alchemy to the Atomic Age* (New York: Plenum, 1995), pp. 205–208.
2. C. L. Cobb, *Magick, Mayhem, and Mavericks: The Spirited History of Physical Chemistry* (Amherst, NY: Prometheus Books, 2002), pp. 10, 250,

252–56, 293, 302, 303, 314, 319, 333, 342, 369, 374, and opening quote; Cobb and Goldwhite, *Creations of Fire*, pp. 310–13, 329, 332, 367, and 401.

## A Whole New Phase

1. C. L. Cobb, *Magick, Mayhem, and Mavericks: The Spirited History of Physical Chemistry* (Amherst, NY: Prometheus Books, 2002), pp. 177–85.
2. Kerry K. Karukstis and Gerald R. Van Hecke, *Chemistry Connections* (New York: Harcourt, 2000), p. 145.
3. Ira N. Levine, *Physical Chemistry*, 4th ed. (New York: McGraw-Hill, 1995), p. 458.
4. Kenneth F. Kiple and Kriemhild Coneè Ornelas, eds., *The Cambridge World History of Food*, 2 vols. (Cambridge: Cambridge University Press, 2000), 1: 693.
5. Roberta Larson Duyff, *The American Dietetic Association's Complete Food and Nutrition Guide* (Minneapolis: Chronimed, 1996), p. 193.

## Equilibirum—Chemistry's Two-Way Street

1. C. C. Gillispie, ed., *Dictionary of Scientific Biography*, 18 vols. (New York: Scribner, 1970–1990), 8: 116–19.

## Colligative Properties—Strength in Numbers

1. C. C. Gillispie, ed., *Dictionary of Scientific Biography*, 18 vols. (New York: Scribner, 1970–1990), 8: 445–47.

## Simply Organic

1. Dorothy Sayers, *The Documents in the Case* (New York: Avon, 1968), p. 201.

## Chemistry Rocks

1. Edgar Allan Poe, *Complete Stories and Poems of Edgar Allan Poe* (New York: Doubleday, 1966), p. 486.
2. David Wilson, *The New Archaeology* (New York: Alfred Knopf, 1975), pp. 88–89.

## The Body of Chemistry Meets the Chemistry of the Body

1. Elaine Marieb, *Human Anatomy and Physiology*, 3rd ed. (Redwood City, CA: Benjamin/Cummings, 1995), p. 881.

2. Stuart Ira Fox, *Human Physiology* (Chicago: Brown, 1996), pp. 372–73.

## Harry, Hogwarts, and Folk Pharmacopoeias

1. C. L. Cobb and H. Goldwhite, *Creations of Fire: Chemistry's Lively History from Alchemy to the Atomic Age* (New York: Plenum, 1995), p. 290.

2. Ibid., p. 285.

3. J. K. Rowling, *Harry Potter and the Sorcerer's Stone* (New York: Scholastic, 1997).

4. Ibid., p. 138; Suzanne E. Weiss, ed., *Foods that Harm, Foods that Heal* (Pleasantville, NY: Readers Digest, 1997), pp. 218, 306.

5. Rowling, *Harry Potter and the Sorcerer's Stone*, p. 138.

6. Deni Brown, *Encyclopedia of Herbs and Their Uses* (New York: Dorling Kindersley, 1995), p. 243.

7. Rowling, *Harry Potter and the Sorcerer's Stone*, p. 81.

8. Brown, *Encyclopedia of Herbs and Their Uses*, p. 228.

9. J. Mann, *Murder, Magic, and Medicine* (New York: Oxford University Press, 1992), p. 66–95.

10. Rowling, *Harry Potter and the Sorcerer's Stone*, p. 273.

11. Ibid., p. 81.

12. Brown, *Encyclopedia of Herbs and Their Uses*, p. 243.

13. Joel Achenback, "Plants on the Warpath: The Roots of Combat," *National Geographic* (February 2004).

# INDEX